GIN O'CLOCK

A YEAR OF GINSPIRATION

WITH OVER 70 DELICIOUS RECIPES TO GUIDE YOU THROUGH THE SEASONS

HarperCollins*Publishers*
1 London Bridge Street
London SE1 9GF

www.harpercollins.co.uk

First published by HarperCollins*Publishers* 2020

10 9 8 7 6 5 4 3 2 1

The Craft Gin Club asserts the moral right to be identified as the author of this work

A catalogue record of this book is available from the British Library

ISBN 978-0-00-837885-1

Food and drink styling: Rebecca Wilkinson
Prop styling: Agathe Gits

Printed and bound by GPS Group, Slovenia

GIN O'CLOCK

A YEAR OF GINSPIRATION

WITH OVER 70 DELICIOUS RECIPES TO GUIDE YOU THROUGH THE SEASONS

HarperCollins*Publishers*

CONTENTS

INTRODUCTION

Here at Craft Gin Club, we hold one truth close to our hearts: there are fantastic ways to enjoy gin all year round! As the co-founders of the UK's first (and largest!) gin subscription club, we've spent every day since 2015 on a quest to find the very best craft gins and deliver them to gin lovers far and wide. Along the way, we've found some absolute crackers and discovered new and exciting ways to serve our favourite spirit.

While we all love nothing more than an ice-cold gin and tonic during the hottest months of the year (or 'G&T season', as we lovingly call it around here), we've also been amazed at the sheer versatility of gin. From recipes for hot gin in December to party punches perfect for spring, gin is a must-have all year round – and it can even play a part in delicious dinners, cakes and bakes. We know that our tens of thousands of members feel the same. That's why every month, along with a full-sized bottle of craft gin, we send a magazine chock full of new cocktails, food recipes, crafting ideas and gin knowledge.

But the more we learn, the more we want to share, and a monthly magazine just isn't enough! That's why we've compiled this book, so that gin lovers across the world can learn more about this spectacular spirit and even more ways to use it. We have collected our favourite cocktails to make at home, alongside fun tips and tricks to help you mix up masterpieces, host gin-soaked soirées and create unforgettable experiences.

While these cocktails, food recipes and craft projects vary in difficulty, every single one is achievable at home. And, though we've grouped the cocktails according to the seasons or occasions when we like to drink them most, they're just like the gin that goes into them – fabulous all year round!

We hope this book will give you the knowledge, confidence and inspiration to put the craft gins in your collection to good use. Happy reading, gin lover!

Cheers!

Jon & John

Co-founders, Craft Gin Club
craftginclub.co.uk

COCKTAIL DIFFICULTY RATING: AN EXPLANATION

Throughout this book, we've given each cocktail one of three ratings: easy, medium and expert. Here's what each rating means.

EASY

An easy cocktail is made using gin and easy-to-find ingredients, like fruit juices and soda water, which are available from your local corner shop. An easy cocktail will also be a snap to make. These drinks have no more than one or two steps.

MEDIUM

Looking for more of a challenge? A medium cocktail will either be made using ingredients that are harder to track down – you may need to go online or to a speciality shop for some of these amazing liqueurs – or have more than one step to completion.

EXPERT

Calling all master mixologists! Expert cocktails require both special ingredients (you may even need to infuse your own gin!) and some more advanced techniques to complete. These tipples are sure to impress, so don't be afraid to give them a go!

QUICK GUIDE TO GIN STYLES

London Dry, Old Tom . . . what does it all mean? Here's a quick guide to gin styles, and what you can do with them.

LONDON DRY

The most readily identifiable style of gin but also the hardest to produce. In 2008, the EU established several rules that gin producers must follow in order to put this label on their bottles. London Dry gins are juniper-forward (i.e. the flavour of juniper predominates) and must be at least 37.5 per cent ABV (alcohol by volume); crucially, no flavours or colours can be added post-distillation, which means the people at the still need to be skilled enough to produce a delicious liquid without any artificial help. There are also limits to how much sweetener can be added to London Dry gins.

OLD TOM

Much sweeter than Plymouth or London Dry styles, Old Tom is the 'missing link' between dry, modern styles of gin and genever, the Dutch distillate that came before them. Popular in the eighteenth century, the Old Tom style has seen a revival at the hands of bartenders, who like it for the 'sweeter something' the additional herbs and spices give to classic cocktails.

PLYMOUTH

Slightly sweeter than London Dry gins, the juniper flavour in Plymouth gin is also subtler, with greater emphasis placed on 'root' botanicals like liquorice and orris. This style must be made in Plymouth – many distilleries used to operate there, but only one remains.

FLAVOURED

Flavoured gins are dominated by a single flavour, which is often added after the distillation process. Raspberry, rhubarb, sloe and elderflower are all popular flavours. Often, the liquid itself will take on the colour of the ingredient flavouring it. To have a go at making flavoured gin yourself, turn to page 116.

MODERN

A modern gin is one that doesn't play by any rules. This catch-all term encompasses anything that doesn't fit into other categories and gives distillers lots of room to manoeuvre. While there are no quality guarantees with modern gins, this category is where some true visionaries are playing – and they have a whole world of flavour at their fingertips!

HOW TO . . .
MAKE A SIMPLE
SYRUP

Making syrups – whether plain and simple or in a range of fantastic flavours – is a very straightforward skill that unlocks a whole world of amazing cocktails.

WHAT IS SIMPLE SYRUP?

Simple syrup is a cocktail mainstay, making an appearance on many an ingredients list. These include the French 75, Bramble and the beloved Tom Collins (see pages 144, 93 and 76). But what is it, and how on earth do you make it?! Simple syrup is – put, well, simply – a combination of sugar and water, heated until the sugar has dissolved evenly and the texture is thick. Also known as sugar syrup, it's used in cocktails to balance sharp and sour flavours.

HOW DO YOU MAKE IT?

It couldn't be simpler: just combine equal parts of caster sugar and water in your saucepan and bring to the boil. Turn off the heat and stir constantly until the sugar has dissolved and the mixture has reduced to a syrupy consistency. Unlike caramel, it's fine to stir the syrup as it cooks. Remove from the heat and allow it to cool in the pan. Then decant it into a clean, sealable container and store in the fridge for up to 2 weeks.

HOW DO I MAKE FLAVOURED SYRUPS?

Just add a bit of what you fancy into your basic syrup! Whether you're using lavender, fruit, spices or veg, letting your flavouring cook alongside your syrup will infuse the finished product. Strain before bottling to remove any bits. Try out as many crazy flavour combinations as you can think of. You may hit upon a new classic!

Here's how to make a few of our favourites . . .

BERRY SYRUP

- 250g (9oz) caster sugar
- 250ml (9fl oz) water
- 100g (3½oz) fresh berries (raspberries, strawberries, blueberries, blackberries or a mixture)

Add the sugar, water and your choice of berries to a saucepan over a medium heat. Simmer, stirring gently, until the fruit is incorporated into the liquid and you've achieved a syrupy consistency. Strain and leave to cool.

LAVENDER HONEY SYRUP

- 60ml (2fl oz) water
- 300ml (10½fl oz) runny honey
- A few sprigs of fresh or dried lavender

Add the water and lavender to a saucepan over a medium heat and simmer gently for 15 minutes. Strain out the lavender and return the infused liquid back to the pan. Bring to the boil then turn off the heat and add the honey. Stir constantly until combined and set aside to cool.

GRENADINE SYRUP

- 200g (7oz) caster sugar
- 250ml (9fl oz) pomegranate juice
- 2–3 drops of orange blossom water (optional)

Add all the ingredients to a saucepan over a medium heat and simmer until the sugar has dissolved and the mixture is syrupy in consistency. Set aside to cool.

HONEY SYRUP

- 250ml (9fl oz) runny honey
- 60ml (1¾fl oz) boiling water

Place the honey in a jar or bowl and pour in the boiling water. Stir constantly until combined and set aside to cool.

GINGER SYRUP

- 150ml (5fl oz) water
- 120g (4½oz) fresh root ginger, sliced
- 150g (5oz) caster sugar

Add the water and ginger to a saucepan over a medium heat and simmer for 30 minutes. Stir in the sugar until syrupy in consistency. Strain and leave to cool.

ROSEMARY AND THYME SYRUP

- 250ml (9fl oz) water
- 2 sprigs each of rosemary and thyme
- 250g (9fl oz) caster sugar

Add the water and herbs to a saucepan over a medium heat and simmer gently for 15 minutes. Add the sugar and stir constantly until dissolved. Strain and leave to cool.

TEA SYRUP

- 2 teabags (such as Earl Grey, English breakfast or chamomile)
- 250ml (9fl oz) boiling water
- 250g (9fl oz) caster sugar

Place the teabags in a saucepan, add the boiling water and leave to steep for a few minutes before removing the teabags. Add the sugar and simmer over a medium heat until thick and syrupy, then set aside to cool.

MUST-HAVE COCKTAIL KIT

We pride ourselves on our cocktail recipes being super-easy to make at home. But, while the blowtorches and centrifuges may be excess to requirements, you'll still need a few basic pieces of kit . . . or, at the very least, some handy substitutes!

COCKTAIL SHAKER

Be it a Boston shaker (see page 42) or a three-piece set, a cocktail shaker is the single most important piece of kit in your collection. You can use these to shake and stir, making them super-versatile.

Or you could use . . . a protein shaker or a plastic water bottle with a watertight lid. Anything that will let you give your liquids a good shake with ice will do!

SPIRIT MEASURE

Most cocktails will call for measures in neat quantities of 25ml (1fl oz) or 50ml (1¾fl oz). A spirit measure is the fastest, easiest way to accurately measure.

Or you could use . . . a measuring cup (for quantities of 100ml/3½fl oz or greater), an egg cup (for measurements of around 25ml/1fl oz), or a tablespoon (15ml/½fl oz) or teaspoon (5ml/¼fl oz) for smaller quantities. These should more or less get you there. If you have absolutely no way of measuring, try counting out your pour. A three-second pour should be about 50ml (1¾fl oz) – but proceed with caution and use common sense!

MIXING GLASS

A glass or metal container used for stirring a cocktail with ice to help cool it down before straining.

Or you could use . . . a measuring jug or indeed any kind of container that you can fill with ice and stir things about in.

JUICER

Psst . . . we've got a secret to share! The single easiest way to take your cocktails from ordinary to extraordinary is using freshly squeezed fruit juice. For this reason, a juicer of some kind is an important piece of kit – unless you fancy doing a lot of squeezing!

Or you could use . . . your hands! This will be harder but needs must. Remember to roll your citrus fruits on the counter or table before slicing in half to release the juices.

BAR SPOON

These long spoons help you to both measure out and stir your cocktail ingredients, without needing to stick your fingers into your drink.

Or you could use . . . a standard teaspoon. This will give you about the same measure as a bar spoon, and a standard spoon will do the mixing just as well.

STRAINER

Whether you're shaking or stirring cocktails, with ice, odds are you'll need to strain them.

Or you could use . . . a spoon! This isn't an exact science (and can get a bit messy!) but if all else fails, hold back the cubes of ice with a spoon while you pour your cocktail into the glass.

SPRING

The sun is shining (sort of!) and the world is coming back to life. These recipes will bring a taste of spring – flowers blooming, bees buzzing and fruit ripening – into your kitchen and gin glass.

APRICOT BLOSSOM

A zingy, springy cocktail to celebrate the
fresh shoots of a new season.

SERVES 1

EASY

- 50ml (1¾fl oz) gin
- 1 heaped tsp apricot jam
- 1 tsp runny honey
- 1 tbsp grapefruit juice
- 1 tbsp fresh lemon juice
- Fresh basil leaves, to garnish

Add your gin to a cocktail shaker. Stir in the jam
and then the honey until both are dissolved. Add
the grapefruit and lemon juices and pack the
shaker with ice. Shake, strain into a cocktail coupe
and garnish with basil leaves to serve.

ENGLISH GARDEN

Step outside – spring is here! This floral refresher is the taste of the world coming back to life.

SERVES 1

EASY

- 50ml (1¾fl oz) gin
- 25ml (1fl oz) elderflower cordial
- 75ml (3fl oz) apple juice
- Slices of cucumber, to garnish

Add the liquid ingredients to a cocktail shaker and fill with ice. Shake and then strain into a highball glass packed with ice. Garnish with slices of cucumber and serve!

FLORAL GIN FIZZES

The classic gin fizz is just four ingredients: gin, lemon juice, simple syrup and soda water. Add in a dash of floral flavour for an easy, refreshing way to welcome spring.

CHAMOMILE GIN FIZZ

SERVES 1

MEDIUM

- 50ml (1¾fl oz) gin
- 25ml (1fl oz) fresh lemon juice
- 3 tsp chamomile syrup (see tea syrup on page 11)
- Soda water, to top up
- Slice of lemon, to garnish

Combine the first three ingredients in a cocktail shaker and fill with ice. Shake and then strain into a highball glass and top up with soda water. Garnish with a slice of lemon and serve.

ELDERFLOWER GIN FIZZ

SERVES 1

EASY

- 50ml (1¾fl oz) gin
- 25ml (1fl oz) elderflower cordial
- Soda water, to top up
- Slice of lemon, to garnish

Fill a highball glass with ice. Combine the gin and elderflower cordial in a cocktail shaker and pack with ice. Shake and then strain into your prepared glass before topping up with soda water. Garnish with a slice of lemon and serve.

BEE'S KNEES

Sweet like honey and tart like lemon, this classic cocktail celebrates the amazing work of our favourite sign of spring: bees! Whether citrus-forward, as in the original recipe, or with a floral twist, this is the perfect tipple for late in the season, when the clocks have gone forward and the evenings are getting ever longer.

CLASSIC BEE'S KNEES

SERVES 1

EASY

- 50ml (1¾fl oz) gin
- 2 tsp runny honey
- 20ml (¾fl oz) fresh lemon juice
- 10ml (½fl oz) fresh orange juice
- Twist of orange peel, to garnish

Add the gin to a cocktail shaker and stir in the honey. Add the lemon and orange juices and fill with ice. Shake well and strain into a cocktail coupe, then garnish with a twist of orange peel.

LAVENDER BEE'S KNEES

SERVES 1

MEDIUM

- 50ml (1¾fl oz) gin
- 20ml (¾fl oz) fresh lemon juice
- 20ml (¾fl oz) lavender honey syrup (see page 11)

Add all the ingredients to a cocktail shaker and fill with ice. Shake, then strain into a cocktail coupe and enjoy!

RHUBARB SOUR

The sour is a classic cocktail format involving a spirit, some lemon or lime juice, something sweet and sometimes also an egg white. This version is a little sweet, a little tart and wholly delicious.

SERVES 1
MEDIUM

- 50ml (1¾fl oz) gin
- 20ml (¾fl oz) fresh lemon juice
- 1 egg white (optional)
- Twist of rhubarb, to garnish

FOR THE RHUBARB SYRUP
- 3 stalks of rhubarb, finely chopped
- 200ml (7fl oz) water
- 200g (7oz) caster sugar

Tip the rhubarb into a saucepan, add the water and simmer over a medium-low heat until the rhubarb is mushy and the water is pink. Strain out the rhubarb pieces and add the sugar. Simmer, stirring, until syrupy in consistency. Cool before using and store in the fridge (see also page 11).

Add the gin to a cocktail shaker with the lemon juice, 20ml (¾fl oz) of rhubarb syrup and the egg white (if using) and shake once, hard, without ice to emulsify the egg. Add ice, shake again and strain into a cocktail coupe. Garnish with a rhubarb twist and serve.

LAVENDER LEMONADE

A refreshing long drink with a beautiful balance of tart, sweet and delicate floral notes, this cocktail is sure to put a spring in your step!

SERVES 1

EASY

- 50ml (1¾fl oz) gin
- 2 tsp runny honey
- Cloudy lemonade, to top up
- Lavender flowers (fresh or dried) and slices of lemon, to garnish

Put the gin into a collins glass and stir in the honey until it's dissolved. Add ice (lavender ice cubes are beautiful in this drink – see page 71) then top up with cloudy lemonade. Garnish with lavender and slices of lemon.

LAVENDER HONEY CHEESECAKE

This light and oh-so-easy cheesecake can be thrown together on a lazy Sunday afternoon, and pairs particularly well with more floral gins. Blue skies or grey, this cheesecake's delicate purple hue is guaranteed to bring a bit of sweet springtime to your plate!

SERVES 8

- 180g (6oz) digestive biscuits, crumbled
- 80g (3oz) butter, softened
- 2 tbsp caster sugar
- 300g (11oz) full-fat cream cheese
- 200g (7oz) mascarpone
- 100g (3½oz) runny honey, plus extra to drizzle
- Violet food colouring gel (optional)
- 1 tsp vanilla extract
- ½ tsp lavender essence
- 2 tsp fresh lemon juice
- 300ml (10½fl oz) double cream
- Culinary lavender grains and honeycomb, to decorate (optional)

YOU WILL ALSO NEED

- 20cm (8in) springform tin
- Piping bag (optional)

1. Preheat the oven to 200°C/180°C fan/Gas 6.
2. Place the biscuits and sugar in a food processor and pulse into fine crumbs, or crush in a resealable plastic bag with a rolling pin. Melt the butter in a saucepan, then remove from the heat and stir in the biscuit crumbs to form clumps.
3. Remove the base of the springform tin and reinsert it so the ridges curve downwards, then press the biscuit mixture firmly into the base. Bake in the oven for 10 minutes, or until crisp.
4. In the food processor or by hand, beat together the cream cheese, mascarpone and honey. If you're colouring the cheesecake, then slowly tint the mixture until the desired shade is reached (see tip).
5. Add the vanilla extract, lavender essence and lemon juice. Pour the double cream into a separate bowl, add a touch more food colouring and whip until thick. Fold this into the cheesecake mixture.
6. Spoon or pipe the mixture on to your biscuit base, then chill the cheesecake in the fridge for at least 6 hours.
7. Serve drizzled with honey and a scattering of lavender, or crumble honeycomb on top for a decadent touch.
8. Enjoy with a Bee's Knees (see page 21) for a truly perfect pairing!

TIP: Dip a toothpick into the food colouring gel and then swirl it into the mixture; repeat and then see if you need to add any more.

HOW TO . . . GROW YOUR OWN GARNISHES

Whether you're working with a windowsill or an allotment, everyone can grow their own gin garnishes . . . and spring is the perfect time to start! Here are the best garnishes to grow for every space, and a few tips to help you get started.

THE WINDOWSILL

Not everyone has acres of space for their horticultural exploration, so the windowsill is an ideal alternative for those who still want to flex their green fingers.

Soft herbs like basil and parsley do well on an outside sill through the summer, but they need to be inside for the winter months. They love to be watered regularly, but basil doesn't like to be damp overnight, so watering it in the morning is best. It is also good to pinch off and discard any flower buds that may appear, so that the leaves can grow nice and bushy!

Chillies are also great for the windowsill, but they do prefer to be indoors. They need lots of light and to be watered regularly, especially when it's hot outside. Aphids can be a problem for chillies, but they can be stopped by spraying the plant with a drop of dish soap dissolved in water.

THE GARDEN

If you do have a garden, evergreen herbs like rosemary, thyme and lavender will do beautifully out there! They are just as happy in a plant pot as a flowerbed, so long as the soil is well drained – they hate soggy roots. The only other thing to remember is to trim them after flowering. Does it get any easier than that?

Mint also does well in the garden, especially as it likes plenty of water. It will spread if given the chance, so unless you want your flowerbeds swamped with the stuff, it might be best to plant it in a spacious pot instead.

THE ALLOTMENT

If your outdoor space isn't big enough for your green-fingered ambitions, allotments offer a great extension for delicious garnish growing. Planting raspberries is a great way to take advantage of the extra space allotments offer. They need a sunny spot and like to be watered regularly in soil that drains well. You will need to construct a wire fence to help them grow straight, but they're well worth the effort.

If you also have a greenhouse, small lemon trees would do very well in a nice pot, so long as they're given citrus plant feed regularly and lots of water in the summer. Be careful in winter, though, as in colder seasons it's best to wait for the top soil to dry out between watering. They also like humidity, so sitting the pot on gravel, on a tray that has been filled with water to the top of the gravel, is a great way to achieve an ideal environment for them.

Happy gardening!

CUCUMBER MARTINI

Mum will love this refreshing twist on a ginny classic.
Best served before a long Mothering Sunday lunch.

SERVES 1
EASY

- 60ml (2fl oz) gin
- 1 tsp simple syrup (see page 10)
- 5 thin slices of cucumber, plus extra to garnish
- Leaves of ½ bunch of mint
- 1 tbsp dry vermouth

Add all your ingredients to a cocktail shaker, then fill with ice and shake. Strain into a chilled Martini glass or cocktail coupe and garnish with a thin slice of cucumber, or a cucumber ribbon (see page 70), to serve.

BREAKFAST MARTINI

On Mother's Day, this is the only thing the special lady in your life will want to see on her breakfast-in-bed tray!

SERVES 1

MEDIUM

- 50ml (1¾fl oz) gin
- 2 heaped bar spoons (or teaspoons) of marmalade
- 1 tbsp triple sec
- 1 tbsp fresh lemon juice
- Orange peel, to garnish

Stir the gin and marmalade together in a cocktail shaker until the marmalade has dissolved, then add the triple sec and lemon juice and fill with ice. Shake for 20 seconds, then strain into a Martini glass. Garnish with orange peel and serve.

HOW TO . . . HOST A GIN TASTING

As a gin lover, you've probably learned lots of useful titbits about your favourite tipple. But a more structured gin tasting is an excellent way to expand your understanding of flavour profiles: how to identify them, enjoy them, mix them, and – most importantly – to find out which gins you like the best. Here's our guide to hosting your own gin-tasting event at home. Bottoms up!

WHAT DO I NEED?

Apart from a selection of craft gins (obvs – more about this later!), get prepared with a few bits of kit before you start:

- A glass per person
- Lots and lots of ice
- Tonics and garnishes
- Spittoon
- Bottle/jug of water (to cleanse the palate between tastings)
- Tasting sheet and score cards (you can find these online)
- Cocktail shaker (if you're going to mix up some masterpieces after the tasting)

HOW MANY GINS SHOULD I TASTE?

How long is a piece of string? How many gins you taste is completely up to you, but we suggest anywhere between three and five. This is enough gins to really note the differences between them, but not so many that you, ahem, lose focus.

Try to find gins that contain different botanicals, are made by different processes or are from different countries. This should give you an interesting evening of tasting completely different gins, will show you the huge variety of flavours available, and help you to discover a new favourite or two!

Work your way through the lot, comparing notes with your gin pals on the different aromas and tastes you can identify.

This is all about having fun as well as learning, so you can score the gins on anything you want, from the originality and uniqueness of the flavour to the attractiveness of the bottle.

HOW DO WE TASTE AND ASSESS THE GIN?

Start by looking at your gin in the glass. Is it clear, or are there particles floating around? (A gin should be clear: it's very rare that it would contain particles that were meant to be there, an example being, perhaps, a novelty gin with glitter in it.) Next, smell the gin, then try a bit of the spirit neat, noting the taste and texture when it hits your mouth and whether there are any secondary flavours coming through. Try to identify the flavours you're tasting, and think about which botanicals might have been used in the distillation process.

If you like the taste of a gin neat, that means you'll probably like it as a G&T, so it's worth experimenting with it further and trying it mixed with tonic water and other ingredients. Tonic and garnishes will change, enhance or mask flavours in your gin, so take note of how the spirit changes.

TIME FOR COCKTAILS?

A gin tasting is all about the different individual flavours of the gins, but you could always mix it up a little (quite literally) by adding a cocktail element to the tasting. Serving a signature cocktail for each gin, designed around the dominant flavours, is a great way to end the evening.

GRAPEFRUIT WHEEL PUNCH

This cocktail is great when you're hosting a crowd, be it for an Easter celebration or a bank holiday bash.

SERVES 6

EASY

- 150ml (5fl oz) gin
- 360ml (12fl oz) grapefruit juice
- 480ml (17fl oz) sparkling white wine
- 3–4 grapefruits

Add heaps of cubed ice to a large punch bowl or jug. Pour over your gin, grapefruit juice and sparkling wine and stir well. Slice the grapefruit into juicy wheels and add them to your punch.

WHITE CHOCOLATE MARTINI

While the kids are enjoying their Easter eggs, the adults can indulge in a very different chocolate treat!

SERVES 1

MEDIUM

- 50ml (1¾fl oz) gin
- 2 tsp crème de cacao blanc
- 2 tsp runny honey
- 1 egg white
- Mini chocolate eggs (unwrapped if covered in foil), to garnish

Add the first three ingredients to a cocktail shaker, stirring until the honey has dissolved. Add the egg white and shake once, without ice, then add ice and shake again. Strain into a cocktail coupe and garnish with mini chocolate eggs.

AVIATION

This lovely, lavender–hued cocktail is beloved by both true gin fans and those new to the spirit – oh, and it packs a punch, to boot!

SERVES 1
MEDIUM

- 45ml (1½fl oz) gin
- 2 tsp Maraschino liqueur
- 1 tsp crème de violette
- 1 tbsp fresh lemon juice
- Maraschino cherry, to garnish

Add all the ingredients to a cocktail shaker, pack with ice and shake. Strain into a cocktail coupe and garnish with a Maraschino cherry. Lovely!

WHITE NEGRONI

With a more delicate, floral flavour than the standard Negroni, this cocktail is a classic Italian aperitivo. Perfect for sipping while you watch the flowers bloom!

SERVES 1

MEDIUM

- 30ml (1fl oz) gin
- 30ml (1fl oz) dry vermouth
- 30ml (1fl oz) Suze
- Twist of lemon peel, to garnish

Pour all the ingredients into a rocks glass filled with ice and stir. Top with a twist of lemon peel to garnish.

HOW TO . . .
USE A BOSTON
SHAKER

They're the two-part cocktail shaker with a penchant for exploding all over the kitchen . . . so why do professional bartenders prefer a Boston shaker, and how do you use one without ruining your kitchen countertops?

WHAT IS A BOSTON SHAKER?

A cocktail shaker comprising two parts: a mixing glass and a metal tumbler, or two metal tumblers. Often the set will include a cocktail strainer, too.

WHY DO THE PROS LOVE THEM?

Simple: they're easier to clean and can hold more liquid – which means more cocktails at once. When you're turning out dozens of drinks an hour, every second counts!

HOW DO YOU USE ONE?

The key to using a Boston shaker successfully is ensuring the vacuum seal between the two parts is perfectly in place. Don't try inserting one half straight into the other; you want them to meet at a slight angle.

Having fitted them together (with your cocktail ingredients inside!), set one half of the shaker down on your work surface and smack the base of the upturned tumbler hard with the heel of your hand. Then put one hand on the seam where the two tumblers meet (make sure you grip hard) and support the top of the upturned tumbler with your other hand. Shake away, making sure that the ice and liquid travel all the way to the top of the upper tumbler!

To open the shaker, give the seam a little crack on the edge of the counter or table to release the vacuum seal. Strain and then enjoy your perfectly shaken cocktail!

ROSE AND CARDAMOM CAKE

Gently spiced and sweetly floral, this cake would be perfect for a Mother's Day tea party.

SERVES 12

- 150g (5oz) butter, softened, plus extra to grease
- 200g (7oz) caster sugar
- 3 eggs
- 2 tbsp rose water
- 1 tsp vanilla extract
- Grated zest of 1 lemon
- 100g (3½oz) plain flour
- 2 tsp ground cardamom
- ½ tsp ground nutmeg
- 2 tsp baking powder
- ½ tsp salt
- 200g (7oz) ground almonds
- 120ml (4fl oz) whole milk
- Dried rose petals and chopped pistachios, to decorate

FOR THE DRIZZLE

- 75ml (3fl oz) gin
- 1 tsp rose water
- 100g (3½oz) sugar

FOR THE GLAZE

- 2 tsp fresh lemon juice
- 150g (5oz) icing sugar
- 1 tbsp whole milk

YOU WILL ALSO NEED

- 20cm (8in) deep-sided cake tin

1. Preheat the oven to 180°C/160°C fan/Gas 4 and grease the cake tin with butter.
2. Using an electric whisk or by hand with a wooden spoon, beat the butter and sugar until pale and fluffy, then slowly beat in the eggs, one by one, until the mixture is smooth.
3. Gently mix in the rose water, vanilla extract and lemon zest. In a separate bowl, sift together the flour, cardamom, nutmeg, baking powder and salt. Add the ground almonds and mix in well.
4. Slowly incorporate the dry mixture and the milk into the batter, alternating between the two.
5. Transfer the batter to your greased cake tin and bake in the oven for 40–50 minutes until the top is a golden brown and a skewer inserted into the middle comes out clean.
6. To make the drizzle, place 50ml (1¾fl oz) of the gin in a saucepan with the rose water and sugar and gently heat until the sugar has dissolved. Remove from the heat and stir in the remaining gin.
7. Score the top of the cake with a skewer and pour the drizzle over the warm sponge. Leave to cool in the tin for 10 minutes, then transfer from the tin to a wire rack.
8. Sift the icing sugar into a bowl, add the lemon juice then slowly add as much milk as needed to form a thick glaze. Pour across the top of the cooled cake. Decorate with dried rose petals and chopped pistachios.

GIN AND LEMON ROAST CHICKEN

This lemony chicken uses common gin botanicals for a light, springtime take on a classic roast that happens to pair perfectly with our favourite spirit.

SERVES 4

- 1 × 2kg (4lb 4oz) free-range chicken, jointed into 4 portions
- 1 lemon, cut into quarters
- Crisp salad or spring vegetables, to serve

FOR THE GLAZE

- Juice and grated zest of 1 lemon
- 60ml (2fl oz) gin
- 100g (3½oz) soft light brown sugar
- 4 juniper berries, crushed
- 1 heaped tsp finely chopped rosemary leaves
- 50ml (1¾fl oz) olive oil
- Salt and freshly ground black pepper

1. Preheat the oven to 220°C/200°C fan/Gas 7.
2. Mix together all the ingredients for the glaze and season with salt and pepper. Line a roasting tin with foil (the sugar can make it a bit sticky!) and place the chicken pieces on top.
3. Spread the glaze over the chicken, pop the lemon quarters around the chicken and roast in the oven for 35–40 minutes, depending on the size of the chicken pieces (see also tip), until the skin is crisp and the juices run clear. Check after 20 minutes or so and if the sugar is browning too much, pop a sheet of foil over the chicken.
4. Once cooked, remove from the oven and leave to rest for 10 minutes before serving with the roasted lemon quarters and a crisp salad or selection of spring vegetables.

TIP: Alternatively, you could roast the chicken whole for around 1 hour 50 minutes.

THE DISTILLER'S GUIDE TO . . . DISTILLATION

It's the process every gin goes through, but what exactly is distillation? Here William Lowe of award-winning Cambridge Distillery explains.

CAN YOU EXPLAIN, AT ITS MOST BASIC, WHAT THE PROCESS OF DISTILLATION IS?

Distillation is the separation and concentration of alcohol and flavours through a process of heating and cooling.

HOW DOES GIN DISTILLATION IN A CLASSIC COPPER STILL WORK? CAN YOU TELL ME, IN SIMPLE TERMS, ABOUT THE SCIENCE BEHIND THE PROCESS?

Gin is a flavoured spirit: for the highest-quality gins, this flavouring is achieved by redistilling the spirit in the presence of botanicals to extract and absorb their essential oils. A base spirit (not 100 per cent alcohol) is poured into the pot of the still, then juniper (and other botanicals) are added. The still is then heated to make the liquid boil. Since the ethanol in the base spirit has a lower boiling point than water, it can be separated by evaporation from the source liquid, the resulting vapour being more highly concentrated in alcohol. This vapour then passes through a condensing coil, where it will come into contact with the cold walls of the coil, turning back into a liquid.

WHAT ARE THE STEPS INVOLVED IN CLASSIC DISTILLATION?

Following the basic methods I've just outlined, the most important steps to consider are the cut points – the boundary points in the process at which the resulting distilled liquid begins to change its character and at which you begin to collect and keep the distilled liquid as it comes out of the still. The liquid comes out in a steady flow; the cut points are carefully timed moments where you switch from collecting the liquid to discard later to collecting the liquid (in a separate container) to dilute with water and then bottle.

Lighter, more volatile elements such as citrus need less energy to evaporate, so they pass through the still early on. Heavier, woodier flavours require more heat to become airborne, so they come through later. The sweet spot, in the middle, is referred to as the 'heart' – the liquid that you collect to use. Even if you start with the same botanicals, changing the time at which you make these cuts in the distillation process will result in totally different styles of gin.

As a very rough rule of thumb, having earlier cuts for the heart (i.e. collecting liquid from an earlier part of the distillation process) can make the gin more floral, lighter and 'lifted' – whereas later cuts can produce earthier, spicier and more robust styles. It is also important to note that being greedy with the cuts (collecting more liquid than you should, with early starting points and later stopping points) may result in a gin that's lower in quality and which may turn cloudy either in the bottle or when diluted: this is known as louching.

WHAT ARE SOME OF THE ESSENTIAL PIECES OF EQUIPMENT USED IN DISTILLATION?

My most used pieces of equipment are surprisingly basic: a watch, a set of scales and a thermometer. I record every aspect of our distillation to ensure everything is completely replicable to avoid unintentional variances in future batches. In terms of more specialist equipment, a density meter is an absolutely essential piece of kit: it's a modern, digital replacement for the hydrometer which enables you to almost instantly determine the ABV (alcohol by volume) of a spirit to two decimal places.

HOW DOES THE TECHNIQUE A DISTILLER MIGHT CHOOSE VARY DEPENDING ON THE BOTANICALS THAT HAVE BEEN SELECTED?

I'd turn this question on its head: up to now, it's been more a question of how a distiller varies the botanicals they're using to suit their chosen distillation techniques. Until very recently, copper-pot distillation was pretty much the default setting. It was brands such as Oxley, Sacred and our own Cambridge Distillery that really pioneered the array of different techniques currently in use.

With those changes now making their way into the wider industry, vacuum distillation is quickly becoming the favoured technique for more delicate, floral botanicals – not least because it permits distillation at much lower temperatures.

HOW HAS CAMBRIDGE DISTILLERY CHANGED THE DISTILLATION GAME? WHAT NEW TECHNIQUES AND EQUIPMENT HAVE YOU USED TO PUSH DISTILLATION FORWARD?

Cambridge Distillery changed the game in three key ways. Firstly, we change our systems to suit our botanicals, not the other way around. For us, flavour comes first, and we won't let practical concerns stand in the way of our pursuit of perfection.

Secondly, we focus on provenance in a way that other distillers simply can't as they have to rely on dried botanicals that are imported. Our gins are not simply locally assembled products, but have true provenance – using botanicals collected from the land directly around the distillery, giving them a terroir, and ensuring that they reflect the spirit of the place they come from.

And last, but certainly not least, we have a deep understanding of how flavour works – not just at a molecular level, but also psychologically. This has enabled us to pioneer our Gin Tailoring techniques to help create a specific gin for a particular person, restaurant or institution – from Michelin-starred restaurants to the House of Lords and the British Airways Concorde Room, to name but a few.

A measure of our success in combining these three elements is that we have been named on three consecutive occasions as the most innovative distillery in the world.

SUMMER

Whether you're having fun at a village fête, heading to the beach or slaving over the barbecue, these recipes will bring the sunniest season to life.

LEMON DRIZZLE SLING

What's even better for a summer picnic than a lovely slice of lemon drizzle cake? A ginny Lemon Drizzle Sling, of course! And you could always still have a slice of cake alongside too . . .

SERVES 1

EASY

- 50ml (1¾fl oz) gin
- 1 heaped tsp lemon curd
- 60ml (2fl oz) pineapple juice
- Juice of ½ lime, plus slices of lime to garnish
- Soda water, to top up

Add your gin to a cocktail shaker and stir in the lemon curd. Add the pineapple juice, lime juice and some ice. Shake and strain into a highball glass filled with ice. Top up with soda water (see tip) and garnish with lime slices.

TIP: You can also serve this cocktail short! Just shake and strain into a cocktail coupe and skip the soda.

BAKEWELL FLUTE

This fancy fizz, inspired by one of our favourite desserts, is a summer celebration in a glass. Perfect for posh picnics!

SERVES 1
MEDIUM

- 5 fresh raspberries
- 35ml (1¼fl oz) gin
- 1 tbsp Maraschino liqueur
- Dash of almond syrup (shop-bought)
- Sparkling white wine, to top up

Muddle three of the raspberries in a mixing glass (see page 113 for the technique), then add your gin, Maraschino liqueur and almond syrup. Top up with ice and mix thoroughly until well chilled. Strain into a champagne flute and top up with sparkling wine. Garnish with the two remaining raspberries and serve.

SUMMER G&TS

Add a fruity twist to the quintessential gin drink with these summery recipes!

STRAWBERRY G&T

SERVES 1

MEDIUM

- 50ml (1¾fl oz) gin
- 1 tbsp strawberry syrup (see berry syrup on page 11)
- Tonic water, to top up
- Fresh strawberries, hulled and sliced, to garnish

Pack a copa de balón glass with ice. Add your gin and the strawberry syrup, then top up with tonic water. Garnish with fresh strawberries and enjoy!

CHERRY G&T

SERVES 1

MEDIUM

- 50ml (1¾fl oz) gin
- Tonic water, to top up

FOR THE CHERRY SYRUP

- 250g (9oz) caster sugar
- 250ml (9fl oz) water
- 100g (3½oz) fresh cherries, plus an extra cherry to garnish

To make the cherry syrup, combine all the ingredients in a saucepan and simmer, stirring, until the cherries have broken down and you have a syrupy consistency. Strain and allow to cool before storing in the fridge.

Fill a copa de balón glass with ice, add your gin and 1 tablespoon of the cherry syrup and stir. Top up with tonic water and garnish with a cherry.

BEACH BEAUTIES

These ginny twists on summer holiday classics are perfect for cooling down in style.

FROZEN GIN AND STRAWBERRY DAIQUIRI

SERVES 1

EASY

- 250g (9oz) fresh strawberries, hulled, plus an extra strawberry to garnish
- 50ml (1¾fl oz) gin
- Juice of ½ lime, plus a slice to garnish
- 1 tbsp simple syrup (optional – see page 10)

Place the strawberries, gin and lime juice in a blender or food processor with 100g (3½oz) of crushed ice (or slightly more ice for a thicker-blended drink) and the simple syrup (if using, though the sweetness of the strawberries should be sufficient). Whizz until smooth and then pour into a hurricane glass and top with more crushed ice. Garnish with a strawberry and a slice of lime.

GIN-A-COLADA

SERVES 1

EASY

- 50ml (1¾fl oz) gin
- 25ml (1fl oz) triple sec
- 25ml (1fl oz) coconut cream
- 1 tbsp ginger syrup (see page 11)
- Juice of ½ lime
- 50ml (1¾fl oz) pineapple juice
- Slice of pineapple and a lime wedge, to garnish

Fill a cocktail shaker with ice and add all of your ingredients. Shake hard and then strain into a cocktail coupe, garnishing with a slice of pineapple and a wedge of lime. If you prefer your drinks blended, chuck all of your ingredients into a blender with ice and blitz before serving in a hurricane glass.

PERFECTLY PINK
SUMMER PUNCH

Gin, rosé and strawberries – this is the stuff summer
is made of!

SERVES 6

EASY

- About 500g (1lb 2oz) fresh strawberries, hulled and sliced
- 150ml (5fl oz) fresh lemon juice
- 5 lemons, sliced
- 90ml (3¼fl oz) simple syrup (see page 10)
- 210ml (7½fl oz) gin
- 480ml (17fl oz) rosé wine
- 120ml (4fl oz) sparkling mineral water, to top up

Start by making the strawberry lemonade. To do this, place the strawberries, lemon juice and sliced lemons in a punch bowl with the simple syrup. Stir and then leave to infuse. When your guests are on their way, add the gin and wine. Stir and top up with sparkling mineral water (see tip).

TIP: Ice can be added to the punch, but only directly before serving. If the water and wine are cold enough, you may not need it.

THE CRAFT GIN CLUB GUIDE TO . . . HOSTING A CRACKING WORLD GIN DAY PARTY

Forget Christmas – the most wonderful day of our year is 8 June, when we celebrate World Gin Day! Why not introduce the gin lovers in your life to this fantastic holiday by throwing an amazing party? Here are our tips for hosting friends and family on World Gin Day!

AROUND THE WORLD IN 80 GINS

Okay, maybe 80 gins is a bit much . . . but why not give your guests a tipple from every continent, at least? Here at Craft Gin Club we've tried a number of spectacular gins from North America, South America, Australasia, Asia and Europe. If you can't rustle up a bottle from Antarctica, why not include one from the UK and one from Europe?

A G&T – OR THREE!

One of the reasons we love gin so much is how versatile it is. Many people think ice and a slice is as good as a G&T can get, but a build-your-own-gin-and-tonic bar will soon prove them wrong.

Set up a table or bar with glasses, buckets of ice, a range of gins and bowls of different garnishes – try spices, fruits, even edible flowers – along with different flavours or brands of tonic. Then let your guests experiment until they find their new favourite serve.

A SIGNATURE START

Set your World Gin Day party up for success by kicking it off right . . . with a cocktail! Choose a signature cocktail for the event – be it one you make up yourself or one selected from the pages of this book – and offer it to your guests right when they arrive. For World Gin Day we love to stick to the classics, like a Negroni or a Bee's Knees. Either is sure to set the mood.

LET US ENTERTAIN YOU

There are so many amazing ways to celebrate our favourite spirit. Whether you choose to host a gin tasting with an intimate group of friends or throw the doors wide open for a proper knees-up, a few ginny games will never go amiss. Try doing a blind tasting, where your guests guess the botanicals in a particular gin. The person to pick out the most flavours wins. Or you can set up a game of Gin Pong! All you need to do is arrange five solo cups in a pyramid on either end of a table. The first person to land a ping pong ball in all five cups wins – and remember that every time you sink a ball, the opponent has to drink the mini G&T in the cup!

PASSION FRUIT SMASH

This mouth-watering combination of gin, fruit juices and passion fruit syrup, is a sure-fire summer smash-hit!

SERVES 1

EASY

- 50ml (1¾fl oz) gin
- 1 tsp fresh lime juice
- 20ml (¾fl oz) fresh orange juice
- Cucumber ribbon (see page 70) and a mint sprig, to garnish

FOR THE PASSION FRUIT SYRUP
- 250g (9oz) caster sugar
- 250ml (9fl oz) water
- Pulp of 3 passion fruit

To make the passion fruit syrup, combine all in the ingredients for the syrup in a saucepan. Heat gently until you have a syrupy consistency, then strain and cool before storing in the fridge until needed (see also page 10).

Fill a cocktail shaker with ice and add 20ml (¾fl oz) of the passion fruit syrup, together with the gin and lime and orange juices. Shake and then strain into a highball glass packed with ice. Garnish with a cucumber ribbon and a mint sprig to serve in style!

PINK GIN AND STRAWBERRY TIRAMISU

Summer is the season of bursting, ripe red berries, and they're put to spectacular use in this pink gin tiramisu! This recipe is perfect for both sophisticated summer garden parties or relaxed barbecues.

SERVES 6

- 1 egg, separated
- 1½ tbsp icing sugar
- 225g (8oz) mascarpone
- 220ml (8fl oz) double cream, whipped
- 1 tsp vanilla extract
- About 30 sponge fingers
- 150g (5oz) fresh strawberries, hulled and sliced, plus extra whole strawberries to serve

FOR THE STRAWBERRY SAUCE

- 250g (9oz) fresh strawberries, hulled and chopped
- ½ tbsp fresh lemon juice
- 1 tbsp caster sugar
- 50ml (1¾fl oz) water
- 75ml (3fl oz) pink gin

YOU WILL ALSO NEED

- 26 x 17cm (10 x 7in) rectangular serving dish

1. Place all the ingredients for the strawberry sauce, except the gin, in a saucepan and bring to the boil. Reduce the heat and simmer very gently for 5 minutes, then remove from the hob and allow to cool. Add the gin and blitz the sauce in a blender until smooth. Pass through a sieve to remove any seeds and pour into a shallow bowl.

2. Now make the mascarpone mixture. Whisk the egg yolk with the icing sugar until pale and fluffy. Fold the mascarpone into the whipped cream and add the vanilla extract, then fold in the egg yolk and sugar mixture. In a clean bowl and using a clean whisk, whisk the egg white into soft peaks and fold this in too.

3. Next dip half the sponge fingers into the strawberry sauce and place in the serving dish, enough to cover the bottom in a single layer. Now cover with half the mascarpone mix and all the sliced strawberries. Repeat with another layer of dipped sponge fingers (pour any leftover strawberry sauce over these, if you like), then the rest of the mascarpone mix.

4. Chill in the fridge for a minimum of 5 hours, or preferably overnight, then top with whole fresh strawberries to serve.

GINARITA

Who needs to hop on a long-haul flight when you could mix up this margarita-inspired gin cocktail instead? Invented by Craft Gin Club friend Adam, it's dangerously easy to drink! The chilli lime salt is an authentic touch, though you could simply wet the rim of your glass with lime and then dip it in salt.

SERVES 1

EASY

- 35ml (1¼fl oz) gin
- 20ml (¾fl oz) triple sec
- 1 tbsp fresh lime juice
- Slices of cucumber, to garnish

FOR THE CHILLI LIME SALT

- 1 tbsp chilli powder
- 1 tsp grated lime zest
- ½ tsp ground cumin
- ¼ tsp each of cayenne pepper, garlic powder, onion powder, ground coriander and salt
- ⅛ tsp caster sugar

Combine all the ingredients for the chilli lime salt and scatter some into a shallow dish. Wet the rim of a Martini glass and dip it in the mixture.

Fill your rimmed glass and a cocktail shaker with ice. Add the gin, triple sec and lime juice to the shaker, then shake and strain into your glass. Garnish with cucumber slices.

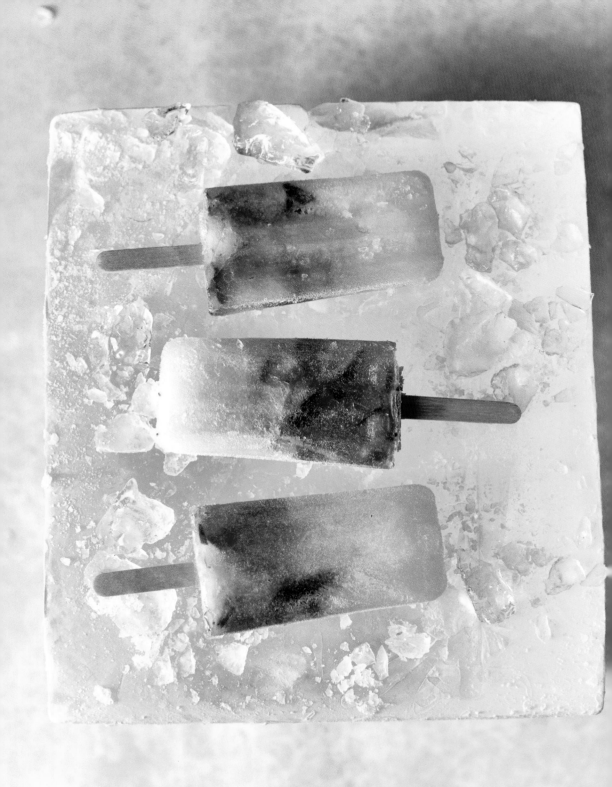

GIN AND TONIC ICE LOLLIES

Why should kids hog all the fun in the sun? When the mercury rises, break out these ice-cold G&T lollies and beat the heat!

MAKES 6 ICE LOLLIES

- 250ml (9fl oz) tonic water
- 75ml (3fl oz) gin
- 50g (2oz) caster sugar
- Sliced strawberries, cucumber slices and fresh mint leaves (or whatever fruits and herbs you like), to garnish

YOU WILL ALSO NEED

- 6-mould ice lolly tray
- 6 wooden ice lolly sticks

Start by making the G&T mix. Open your tonic and let the bubbles dissipate. Then combine in a mixing jug or bowl with the gin and sugar. Stir gently and set aside.

Fill the lolly tray with your strawberries, mint and cucumber – we like to roughly chop everything and sprinkle a small mixed handful into each lolly mould. Pour over your G&T mix and freeze for at least 4 hours, but preferably overnight.

CLOVER CLUB

A classic bursting with the berry flavours of midsummer, this recipe is the original! A crowd-pleaser that is a must-try for every gin lover.

SERVES 1
MEDIUM

- 30ml (1fl oz) gin
- 30ml (1fl oz) dry vermouth (optional)
- 20ml (¾fl oz) fresh lemon juice
- 20ml (¾fl oz) raspberry syrup (see berry syrup on page 11) or raspberry jam
- 1 egg white
- 3 fresh raspberries, speared with a cocktail stick, to garnish

Add all the ingredients to a cocktail shaker and dry shake (without ice). Then add ice and shake for a second time. Strain into a chilled cocktail coupe. Garnish with the speared raspberries.

HOW TO . . . MAKE YOUR OWN GARNISHES

CITRUS PEEL

Using a vegetable peeler or paring knife, cut off the peel from your citrus fruit, focusing on the peel itself – no pith or flesh needed here! These are fabulous in short drinks, such as the Negroni (see page 111).

CITRUS TWIST

Using a vegetable peeler or a paring knife, remove a strip of peel from your citrus fruit (for a longer strip, go widthways around the fruit; for a shorter piece, go from top to bottom). Wrap the peel, pith side down, in a coil around a pen, pencil or chopstick. Hold fast for a moment (20 seconds should be enough) and slide the twist free. Drop into your drink carefully to maintain the twist.

BURNT ORANGE PEEL

Cut a small rectangle of peel from an orange and hold it in one hand. Hold a lighter in your other hand a few centimetres below the piece of peel, making sure that the pith side is facing away from the lighter. Turn on the lighter and squeeze the peel over the flame, spritzing the oils into it, which will toast the peel. This garnish is spectacular in our Craft Gin Club's Christmas Cocktail on page 126 or the Bonfire Bramble on page 88.

CUCUMBER RIBBON

Use a vegetable peeler to cut long, thin strips from a cucumber. Why not use a few in each drink for an extra punch? Cucumber ribbons look great in tall drinks, such as the Passion Fruit Smash (see page 60).

BOTANICAL ICE CUBES

Sprinkle your chosen garnish into an ice-cube tray – we love making ice cubes with lavender grains, pink peppercorns, chamomile blossoms or mixed berries – then fill with water and freeze. When frozen, simply drop into your drink! We love using these in our G&Ts and also in the Lavender Lemonade on page 24.

FRUIT OR FLOWER SKEWER

Using a toothpick or a cocktail skewer, spear your garnishes and either drop them straight into your drink or perch on the rim of the glass. We love using this technique with dried rosebuds, fresh raspberries and even pickled onions! Great for Martinis, a raspberry skewer is a must-have for the Clover Club on page 68.

FRUIT SLICE

Using a sharp vegetable knife, cut your fruit – an apple, lemon, grapefruit or lime or any fruit that has both peel and flesh – into quarters lengthways, and core if needed. Taking one quarter, slice into smaller wedges, making sure that each wedge has some skin and flesh. For a smaller fruit, such as a lime, simply cut in half lengthways before slicing into wedges. Pop right into your drink or freeze to use in place of ice cubes. These are perfect garnishes for G&Ts!

WATERMELON GIN PUNCH

This is the perfect pool-party punch – it's refreshing, colourful and, of course, full of gin!

SERVES 6

EASY

- ½ large watermelon, peeled, cut into chunks and seeds removed, plus extra slices to garnish
- 250ml (9fl oz) gin
- 125ml (4fl oz) fresh lime juice
- 75ml (3fl oz) simple syrup (see page 10)
- Fresh mint leaves, to garnish

Place the watermelon chunks in a food processor or blender and blitz until liquid. Add to a punch bowl filled with ice. Add the gin, lime juice and simple syrup and stir. Garnish with watermelon slices and mint leaves and serve.

GINNY APEROL SPRITZ

As apt in a palazzo as a pub garden, the Aperol spritz is a refreshing, complex treat with a summery orange glow – and it's only improved by adding a measure or two of gin!

SERVES 1.

EASY

- 2 tsp gin
- 25ml (1fl oz) Aperol
- Sparkling white wine or soda water, to top up
- Slice of grapefruit or orange, to garnish

Combine your gin and Aperol with ice in a wine glass. Top up with soda or sparkling wine and stir well. Garnish with a grapefruit or orange slice and serve.

TOM COLLINS

An elegantly simple, stone-cold classic, the Tom Collins dates back to the nineteenth century, but the passing years haven't dimmed the appeal of this ultimate long drink. The perfect summer cooler, it's refreshing, versatile and super-easy to make – whether you opt for the traditional version or our elderflower-infused twist.

CLASSIC TOM COLLINS

SERVES 1

EASY

- 50ml (1¾fl oz) gin
- 25ml (1fl oz) fresh lemon juice
- 20ml (¾fl oz) simple syrup (see page 10)
- Soda water, to top up
- Lemon wedge, to garnish

Fill a highball glass with ice and add your ingredients one at a time. Top up with soda water and garnish with a lemon wedge.

ELDERFLOWER TOM COLLINS

SERVES 1

EASY

- 35ml (1¼fl oz) gin
- 15ml (½fl oz) elderflower liqueur
- 2 tsp simple syrup (see page 10)
- 2 tsp lemon juice
- Soda water, to top up
- Lemon wedge, to garnish

Fill a highball glass with ice and build your drink inside, adding each ingredient one at a time and topping up with soda water. Give it a stir and garnish with a lemon wedge.

HOW TO . . . PERFECTLY GARNISH YOUR G&T

A garnish can enhance some flavours in your gin and mask others. Here's how to pick the perfect garnish for any gin.

COMPLEMENT

If you're just starting on your gin journey, you can take a short cut to garnish perfection. Just take a look on the back of your gin bottle. Choose one of the botanicals used to make the gin you're drinking and select your favourite. Garnishing with a botanical will enhance that flavour in the finished drink.

CONTRAST

Once you get a little more comfortable garnishing with complementary flavours, why not experiment with contrasting flavours? Zero in on the dominant flavour in the gin you're enjoying and pair it with something completely different. For example, does your gin have strong orange notes? Garnish with a cinnamon stick and see how it changes.

COMBINE

Is your garnish game already on point? Start experimenting by using multiple garnishes. We love the combination of citrus and floral – lemon and elderflower is a classic. Fruity with spice is delicious too – how about strawberries and black peppercorns? There's a whole world of garnishes waiting to be explored!

CRAFT GIN CLUB'S FOOLPROOF GARNISH GUIDE

If you want to stick with tried-and-true flavour combinations, try these!

- If your gin is citrus-forward, with flavours of lemon and grapefruit . . . garnish with bergamot, basil, star anise, mint or lemon thyme.
- If your gin is spiced, with flavours of cinnamon, star anise or pepper . . . garnish with black pepper, black tea leaves, ginger, orange, cinnamon or pink peppercorns.
- If your gin is juniper-forward . . . garnish with grapefruit, juniper berries, lemon or lime.
- If your gin is floral, with flavours of rose, chamomile, elderflower or lavender . . . garnish with rosehip, lavender, apple, rosemary, chamomile, rosebuds or rhubarb.
- If your gin is fruity, with flavours of berries or apples . . . garnish with chilli (fresh or dried), basil, berries (fresh or dried), redcurrants, pomegranate or black pepper.

GIN BARBECUE RECIPES

What's summer without a long weekend spent at the barbecue? The recipes here incorporate gin into backyard barbecue favourites – and we're totally obsessed with them! Cook these tasty treats either in the oven or on a barbecue and serve alongside a big bowl of punch or at a 'make your own G&T' bar for a casual, gin-soaked afternoon in the sun.

STICKY GIN AND HONEY BARBECUE CHICKEN WINGS

SERVES 4–6

- 2kg (4lb 6oz) chicken wings
- 1 heaped tbsp sesame seeds, lightly toasted in a dry pan
- Bunch of spring onions, thinly sliced
- Griddled lemon halves, to serve (see tip)

FOR THE MARINADE

- 4 tbsp light soy sauce
- 1 tsp ginger paste or finely grated fresh root ginger
- 1 tsp garlic paste or finely chopped garlic
- 340g (12oz) runny honey
- 120ml (4fl oz) rapeseed oil
- 60ml (2fl oz) gin
- Juice and finely grated zest of 1 lemon
- ½ tsp salt
- ½ tsp finely ground white pepper

1. First prepare the wings. If they still have the wing tips attached, feel free to cut these off and save them to make chicken stock (just pop them in the freezer in a bag). Now, cut each wing in two at the joint.
2. Place all the marinade ingredients into a large resealable plastic bag and squelch around until fully mixed. Add the chicken wings, close the bag and squish everything around again until the wings are completely coated in the marinade. Leave to marinate in the fridge for at least an hour or preferably 3–4 hours.
3. When you are ready to cook, preheat the oven to 220°C/200°C fan/Gas 7. (You could also cook these over a barbecue for 20–30 minutes, turning frequently, or until cooked through.)
4. Spread the wings out in a shallow roasting tin so they are in a single layer, then pour any remaining marinade on top.
5. Roast in the oven for 30–35 minutes, turning the wings and basting them in the juices from time to time during cooking.
6. When almost all the liquid has gone and the wings are golden brown and sticky, remove from the oven, pile on to a serving dish and sprinkle with the sesame seeds and spring onions. Serve with griddled lemon halves and Green Slaw (see page 80).

TIP: To make griddled lemon halves, cut 2–3 lemons in half and place, cut side down, on a griddle or frying pan and cook until golden.

GREEN SLAW

SERVES 6–8

- 40g (1½oz) kale, chopped
- 40g (1½oz) baby spinach
- ½ white cabbage, shredded
- 3 Granny Smith apples (unpeeled), coarsely grated
- 30g (1oz) rocket

FOR THE DRESSING

- 5 tbsp low-fat Greek-style yoghurt
- 5 tbsp mayonnaise
- Juice of 1 lemon
- 1 tsp Dijon mustard
- Freshly ground black pepper

1. Place the kale, spinach, cabbage and apple in a bowl and mix together. Pop the dressing ingredients into an empty screw-top jar, put the lid on and shake well, or add the ingredients to a jug and mix thoroughly.

2. Pour the dressing over the slaw and toss all the ingredients together until well coated. Pile the rocket on top of the dressed slaw to serve.

GIN PULLED PORK

SERVES 6–8

- 150g (5oz) salt
- 300g (11oz) soft light brown sugar
- Pared zest and juice of 1 lemon

1. Make an aromatic brine by placing the salt and half the sugar in a large saucepan with the lemon and lime zest, juniper berries, cardamom pods, teabag, 2 of the bay leaves and the water. Bring to the boil, ensuring the salt and sugar have completely dissolved, then remove from the heat and allow to cool.

- Pared zest and juice of 1 lime
- 6 juniper berries
- 4 green cardamom pods
- 1 green jasmine teabag
- 4 bay leaves
- 1 litre (1¾ pints) water
- 1 x 3kg (6½lb) boned pork shoulder, skin removed
- 50ml (1¾fl oz) gin
- 300ml (10½fl oz) vegetable stock
- 100ml (3½fl oz) cider vinegar
- 2 garlic cloves, crushed
- 100g (3½oz) runny honey

2. Place the pork shoulder in a large dish, pour over the brine, cover in foil or cling film and leave in the fridge for up to 24 hours.
3. When you are ready to cook, preheat the oven to 140°C/120°C fan/Gas 1. (For cooking on a barbecue, see tip.)
4. Remove the pork from the brine and dab dry with kitchen paper. Wrap in foil, making sure the pork is completely sealed, and place in a roasting tin. Slow-cook in the oven for 4–5 hours, until the meat is completely tender and falling apart.
5. Leave the pork to rest, still covered in foil, for 30 minutes. Unwrap and shred using two large forks.
6. Place the remaining sugar and the lemon and lime juices in a large saucepan (large enough for the pork to fit in as well) with the gin, stock, vinegar, garlic, honey and the last 2 bay leaves and slowly bring to the boil.
7. Reduce the heat to a very low simmer, add the shredded pork, mix well and continue to simmer until the liquid has reduced to a sticky glaze. This should take around 40 minutes.
8. Serve with Green Slaw (see opposite) and enjoy!

TIP: To make this on a barbecue, heat up the barbecue to as close to 120°C as possible, putting the meat on the grill alongside a pan of water for keeping the pork moist as it cooks. Add hardwood chunks, topping up every hour after the first 4 hours. When the meat reaches 65°C (on a meat thermometer), wrap in 2 layers of foil. Cook for 6–7 hours, or until the internal temperature is 85°C, then remove from the barbecue and shred as above.

GINOLOGY 101:
THE DISTILLER'S GUIDE TO BOTANICALS

You've probably heard all about botanicals – which ones are in your favourite gin, how they were distilled and where they're from. But what actually is a botanical? Here Cameron Mackenzie of Four Pillars Distillery in the Yarra Valley, Australia – who's made everything from a Bloody Shiraz Gin to a Spiced Negroni – explains all.

WHAT ACTUALLY IS A BOTANICAL?

A botanical is, in my mind, a spice or natural substance that provides aroma and flavour to a distilled spirit.

CAN ANYTHING BE USED AS A BOTANICAL?

Well, yes and no.

I find that the best botanicals are fairly high in oil content. We're using alcohol as a solvent to carry the oils through the still – ultimately this will become gin. So if a botanical is very low in oils, there won't be a great deal of flavour.

In distilled spirits, like gin, you can use any number of botanicals. The key is to make sure they work with juniper, the base botanical for all gins. If a botanical doesn't 'match up' with juniper, it can create a clash of flavours.

EXCLUDING JUNIPER, WHICH MUST BE USED IN GIN, WHAT ARE THE MOST COMMON GIN BOTANICALS?

Coriander seed is likely to be the most-used botanical after juniper. Cardamom, cinnamon, anise and other spices are also very traditional, which reflects the spice trade over the last 200 years – as gin was developing, these ingredients were readily available and their flavours very popular.

As gin has expanded, I think that every gin-making nation has one or two hallmark local botanicals. In Australia we have lemon myrtle, strawberry gum and pepperberry.

WHAT'S THE MOST UNUSUAL BOTANICAL YOU'VE EVER WORKED WITH?

The most unusual to date is parsnip! We've only done some trials, but the results have been amazing, and not what we were expecting at all.

The trickiest, though, was star anise. It has to be used sparingly or everything tastes like ouzo! It's a balancing act.

HOW DOES THE DISTILLING PROCESS CHANGE WHEN YOU'RE USING FRESH BOTANICALS – LIKE FRESH FRUITS OR FLOWERS – INSTEAD OF DRIED ONES?

We tend to put fresh botanicals in an external botanical basket and vapour-infuse them. Personally, I find that the botanical basket doesn't get quite as hot as the distilling kettle, so we can extract the oils without cooking the botanicals. Dried spices hold up to the heat pretty well, so we tend to macerate them.

AS A DISTILLER, WHAT'S YOUR FAVOURITE BOTANICAL?

When we were developing our navy-strength gin, we had a box of what we thought was root ginger arrive, but when we opened it, we found that it was full of fresh turmeric! To this day, turmeric remains my favourite-ever botanical. It has characteristics of dill, cucumber and carrot – very fresh, but also very rich. It's delicious!

A QUICK GUIDE TO GLASSWARE

How do you choose a glass for your fabulous cocktails? This easy guide will help you get it right every time!

CHAMPAGNE FLUTE

Perfect for . . . a French 75 (see page 144), or anything using sparkling wine. With their narrow rims and long stems, flutes keep cocktails perfectly chilled and bubbling for longer.

COCKTAIL COUPE

Perfect for . . . shaken cocktails like the Aviation, Bee's Knees or Clover Club (see pages 40, 21 and 68). A wide rim releases all of the amazing aromas of your shaken cocktail, while the long stem ensures your hand won't heat up your drink. Supremely stylish and tough to spill.

COPA DE BALÓN

Perfect for . . . G&Ts! The capacious bowl allows you to pack in the ice, meaning a less diluted G&T, while the wide rim gives your gin's aromas lots of opportunity to shine.

WINE GLASS

Perfect for . . . a Ginny Aperol Spritz (see page 74) or indeed any shaken cocktail at a pinch! The ubiquitous wine glass is a great all-rounder. With lots of room for ice and garnishes but with a long stem to keep your drink cold, they are a great option for spritzes and, if you don't have a cocktail coupe to hand, a good second choice for serving shaken cocktails.

HIGHBALL

Perfect for . . . a Gin Fizz, Gin Sling or Tom Collins (see pages 20, 50 and 76). These tall, straight glasses are great for drinks with lots of ice and a mixer. They're tall and narrow, which keeps the liquid cold and carbonated.

HURRICANE GLASS

Perfect for . . . tropical tipples like the Frozen Gin and Strawberry Daiquiri or Gin-a-Colada (see page 55). A mainstay at beach bars across the world, hurricane glasses are great for blended drinks. With lots of room for liquid – blended drinks can be bulky! – and creative garnishes, they instantly evoke vaycay vibes.

MARTINI GLASS

Perfect for . . . a Gin Martini or Gimlet (see pages 30, 33, 122, 142 and 106). A long stem keeps drinks served 'straight up' – or without ice – nice and cold. And what a stylish shape!

ROCKS GLASS

Perfect for . . . a Negroni, Bramble or Gin Sour (see pages 111, 93 and 104) – anything that packs a punch! Also called tumblers, old-fashioned glasses or lowballs, the rocks glass is perfect for short, strong drinks. A thick base makes muddling a breeze, so you can 'build' a cocktail in these robust glasses.

AUTUMN

From the hedgerows bursting with ripe berries to the seasonal produce stacked on market stalls, this is the season of plenty. Celebrate the bountiful months of autumn with these flavourful recipes.

BONFIRE BRAMBLE

Combining the succulent berry flavours of autumnal hedgerows with the subtle heat and smoke of seasonal bonfires, this is one tasty twist on a classic cocktail. To try the original in its pure form, check out the Bramble recipe on page 93.

SERVES 1

EASY

- 50ml (1¾fl oz) gin
- 1 heaped tsp blackcurrant jam
- 1 thumb-sized piece of fresh root ginger, peeled and chopped into 3 segments
- 50ml (1¾fl oz) apple juice
- 2 tsp fresh lemon juice
- Grind of black pepper
- Burnt orange peel (see page 70), to garnish

Place the gin in a cocktail shaker. Stir in the jam, then add the ginger, apple juice and lemon juice, plus a grind of black pepper. Fill the cocktail shaker with ice, shake and strain into a cocktail coupe. Garnish with burnt orange peel.

AUTUMNAL SMASH

If you're pulling out the jumpers and boots at the first sign of a chill in the air, these two cocktails are perfect for you! Refreshing enough for the warm days of early September, these smashes still manage to capture the golden glow of autumn.

HEDGEROW SMASH

SERVES 1

EASY

- 2–3 blackberries, plus extra to garnish
- 6–7 fresh mint leaves, plus extra to garnish
- 50ml (1¾fl oz) gin
- 1 tbsp simple syrup (see page 10)
- 1 tbsp fresh lemon juice
- Sparkling mineral water, to top up

In the bottom of a highball glass, muddle the blackberries and mint leaves (see page 113) – it's not called a 'smash' for nothing! Add ice to your glass, then the gin, simple syrup and lemon juice. Give it a good stir, then top up with sparkling mineral water. Garnish with blackberries and mint.

PEAR SMASH

SERVES 1

MEDIUM

- 1 overripe pear, peeled, cut in half and core removed
- 50ml (1¾fl oz) gin
- 1 tbsp simple syrup (see page 10)
- 1 tbsp fresh lemon juice
- Sparkling mineral water, to top up
- Rosemary sprig, to garnish

Place the pear in a bowl and crush using a potato masher. Push the pulp through a sieve into a highball glass, then add ice. Add the gin, simple syrup and lemon juice. Stir, then top up with sparkling mineral water. Garnish with a rosemary sprig and serve.

CREEPY CLASSICS

If you're hosting a ghoulishly glamorous Halloween party, skip the pumpkins and go straight to the dusty tomes of cocktail history. Haunting those pages are the ghostly White Lady and creepy Corpse Reviver No. 2 – both delicious concoctions worth bringing back from the graveyard of mixology.

WHITE LADY

SERVES 1

EASY

- 45ml (1½fl oz) gin
- 25ml (1fl oz) triple sec
- 1 tbsp fresh lemon juice
- 1 egg white

Combine your ingredients in a cocktail shaker and dry shake (without ice) for 20 seconds until the egg is emulsified. Add ice and shake again. Strain into a chilled cocktail coupe and serve straight away.

CORPSE REVIVER NO. 2

SERVES 1

MEDIUM

- 25ml (1fl oz) gin
- 25ml (1fl oz) triple sec
- 25ml (1fl oz) Lillet Blanc
- 25ml (1fl oz) fresh lemon juice
- 1 tsp absinthe
- Lemon peel, to garnish

Combine all the ingredients with ice in a cocktail shaker. Shake and then strain into a chilled cocktail coupe. Garnish with lemon peel.

BRAMBLE

A modern classic, the Bramble perfectly captures that amazing moment when the hedgerows burst into life in the earliest days of autumn.

SERVES 1

MEDIUM

- 60ml (2fl oz) gin
- 30ml (1fl oz) fresh lemon juice
- 2 tsp simple syrup (see page 10)
- 2 tsp crème de mûre (blackberry liqueur, see tip)
- Fresh blackberries, to garnish

Add your gin to a cocktail shaker with the lemon juice and simple syrup. Pack the shaker with ice and shake vigorously, then strain into a glass filled with crushed ice. Drizzle the crème de mûre over the ice and garnish with blackberries to serve.

TIP: If you have a sweet tooth – or can't be bothered to buy a bottle of crème de mûre – swap it for blackberry syrup (see berry syrup on page 11).

HEDGEROW CRUMBLE BARS

Everybody loves a crumble, and these moreish bars elevate the classic recipe into something quite special. To us there's no finer combination than freshly foraged blackberries and juicy autumn apples, but you can easily substitute the fruit here for whatever you have stashed away – rhubarb or plum, for example, would give a juicy bite to the bars. Got last year's sloe gin waiting to be used up? Swap it in for sweetly sticky twist!

**MAKES 16 SQUARES
OR 8 BARS**

FOR THE BLACKBERRIES
- 200g (7oz) blackberries
- 100ml (3½fl oz) gin
- 2 tbsp sugar (omit if using sloe gin)

FOR THE APPLES
- 500g (1lb 2oz) Bramley apples
- 100ml (3½fl oz) water
- 2 tsp ground cinnamon
- 1 tsp ground nutmeg
- 1 tsp vanilla extract
- 50g (1¾oz) soft light brown sugar
- 2 tbsp cornflour

1. To macerate the blackberries, toss them in a container with the sugar and gin. Leave to soak for a few hours, then strain off the flavoured gin and reserve both the macerated fruit and the gin.

2. Preheat the oven to 200°C/180°C fan/Gas 6 and line the cake tin with baking parchment.

3. Peel, core and dice the apples. Add to a medium saucepan with half the water, the cinnamon, nutmeg, vanilla extract and brown sugar and simmer over a low heat for 5 minutes. Combine the remaining water with the cornflour. Add the mixture to the pan, along with the reserved gin, and simmer for another 5 minutes until the apples are soft.

4. Next make the crumble. In a bowl, mix together the sugar, flour, baking powder and oats. Pour in the melted butter and combine until you reach a sandy consistency. Keep a third of the mixture aside for the topping and press the remaining two-thirds of the crumble into the base of your prepared tin. Using a glass with a bit of baking parchment wrapped around the bottom works well here to get a densely packed base.

Continues over the page

FOR THE CRUMBLE

- 200g (7oz) soft light brown sugar
- 220g (8oz) plain flour
- 1 tsp baking powder
- 100g (3½oz) porridge oats
- 200g (7oz) slightly salted butter, melted
- Flaked almonds, to sprinkle

FOR THE CREAM CHEESE FILLING

- 250g (9oz) full-fat cream cheese
- 25g (1oz) icing sugar
- 1 egg
- 1 tsp vanilla extract
- 1 tbsp fresh lemon juice

YOU WILL ALSO NEED

- 20cm (8in) square cake tin

5. To make the cream cheese filling, beat together the cream cheese, icing sugar, egg, vanilla extract and lemon juice with an electric whisk or by hand with a wooden spoon until light and fluffy. Use a spatula to layer this on top of your crumble base.

6. Evenly scatter the macerated blackberries on top of the cream cheese filling, then cover with the cooked apple mixture. Finally, cover with the remaining crumble.

7. Bake on the middle shelf of the oven for 20 minutes. Scatter the top with flaked almonds, then bake for another 10 minutes until the top is golden and crispy. Leave to cool for an hour, then chill in the fridge for a few hours before removing from the tin and slicing into squares.

ASK THE EXPERTS: THE CARE AND KEEPING OF GIN

These are some of the the most common questions we get asked by our members, and the answers every gin lover needs to know.

HOW LONG DOES UNOPENED GIN LAST?

As long as the seal on your gin is unbroken – so no air can get in the bottle – your gin will stay in pristine condition for years. But you shouldn't wait too long! Unlike wine, gin won't get better the longer you leave it in the bottle. It's pointless to wait, hoping your gin will improve with time.

HOW LONG CAN I KEEP A BOTTLE OF GIN ONCE IT'S OPEN? WILL IT EVENTUALLY GO OFF?

The bad news is that your gin will eventually go off once it's opened. The good news is that it takes a seriously long time! As soon as you open a bottle of gin, the oxidisation process starts. This process won't spoil your gin, but it will change the way your gin tastes. After a long period of time exposed to air, all of the nice flavours of your gin will disappear. We recommend finishing an open bottle within a year. But, if you want to slow down the oxidisation process, there are a few things you can do:

1. Let your gin live in the fridge
Cold temperatures slow down the oxidisation process, so if you have room in the fridge or the freezer for your open gin bottles, that's the best place to keep them. Failing that, anywhere with a stable temperature should do – for example, under the bed, in a dark cupboard or on a shaded shelf.

2. Scale down
The less room for air in your gin bottle, the longer the liquid will last. If you've gone through a good chunk of a gin but want to keep hold of the rest, decant it into a smaller bottle and seal.

3. Keep upright
If your gin has a natural cork, be sure to keep it upright. Spirits can eat through the cork, which in turn taint the liquid.

HOW CAN I TELL IF MY GIN IS OFF?

It's very rare for gin to go off, but it's also easy to tell when it's happened. Give your gin a sniff – if it smells strange, or you see any unusual particles floating around, tip it down the sink and open a new bottle.

A DISTILLER'S GUIDE TO . . . GIN STYLES

Stuart Nickerson, director of the Shetland Distillery Company – the man responsible for the amazing range of Shetland Reel gins – explains what a flavour profile is and how it can help you find new favourites.

AT ITS MOST BASIC, WHAT ARE THE FLAVOUR PROFILES THAT A GIN CAN HAVE?

Obviously, the predominant flavour of a gin will be juniper – that's the flavour that literally defines gin. After that, distillers have a complete palette of flavours at their disposal.

In the past, gins have tended to have well-balanced flavours of citrus and spice with juniper, but nowadays you can have so much variation. You can find big citrus gins, very spicy gins, very sweet Old Tom gins or even wood-influenced, cask-aged gins like our Up Helly Aa. The main thing, though, is to always produce a well-balanced spirit that's still a gin – and that means predominant flavours of juniper.

WHAT FACTORS INFLUENCE THE FLAVOUR PROFILE OF A GIN?

The first and most obvious influence is the botanicals used in the distillation process, including the proportions of each botanical in the recipe. Distillation is a lot like cooking – vary the ingredients or their relative proportions and you get a different flavour profile.

The second influence is the method of production. When we're distilling our gins, we can influence the flavours by changing the length of time that we steep or macerate the botanicals in the neutral grain spirit. We steep all of our gins for a minimum of 24 hours, because we believe that it's the best amount of time to optimise the flavour.

Flavour is also influenced by the speed of distillation, and by the cut points used for each distillation. These are settled on after a lot of trial and error on the part of a distiller. Some distilleries will add flavours to their gins after they've been distilled, which changes the flavour again. We haven't ventured down this route.

WHY IS IT IMPORTANT TO KNOW THE FLAVOUR PROFILE OF THE GIN YOU'RE DRINKING?

Each of us has our own preferred flavours, so when we find gins that we like and know their flavour profiles, it helps us when it comes to trying new gins. You'll immediately have a reference point, allowing you to compare new gins you try with the gins you've enjoyed before with the same flavour profile.

WHY IS IT IMPORTANT FOR THE SHETLAND DISTILLERY COMPANY TO OFFER SEVERAL FLAVOUR PROFILES WITHIN YOUR GIN RANGE?

We have a large number of discerning and adventurous customers who like to try new things. For that reason we like to offer a range of products, which include everything from a very citrus-forward spirit like our Filska gin to a spicier gin like Wild Fire, or for the more adventurous, a cask-aged, navy-strength gin: Up Helly Aa.

We're working on new flavour profiles all of the time: all of the development work takes place in my kitchen, with my wife and fellow director Wilma in charge. It means that our house has a lot of samples lying around!

WHAT ARE THE MAIN FLAVOUR PROFILES A GIN CAN HAVE?

Citrus
Examples: grapefruit, orange, lemon

Floral
Examples: lavender, rose, apple or orange blossom

Herbaceous
Examples: sage, rosemary, thyme

Spiced
Examples: cinnamon, star anise, cardamom

Juniper-forward
The piney notes of juniper dominate

HOW CAN I FIGURE OUT WHAT FLAVOUR PROFILE A GIN HAS?

Step One: Identify the botanicals
The botanicals list will give you a sneak peek into how a gin might taste. Think of the flavours as you know them from food to get a sense of how they'll interact.

Step Two: Give it a taste!
Pick out the dominant flavour or two and think of which category they would fall under. That's your flavour profile!

OKTOBERFEST PUNCH

Beer and gin may seem like strange companions, but they can combine to make some cracking cocktails! We love this combination of ale, grapefruit juice and gin – a citrus-forward spirit works best – for the autumnal celebration of Oktoberfest.

SERVES 6–8

MEDIUM

- 120ml (4fl oz) gin
- 60ml (2fl oz) elderflower liqueur
- 480ml (17fl oz) golden or pink grapefruit juice
- 120ml (4fl oz) simple syrup (see page 10)
- 2–3 grapefruits, sliced into wheels, to garnish
- 1 x 350ml (12fl oz) bottle of India pale ale, to serve

In a large jug, combine the gin and elderflower liqueur, then add the grapefruit juice and simple syrup. Grab your cocktail stirrer and mix thoroughly. Toss in a generous portion of ice and stir until you feel the chill as you grasp the jug. Garnish with wheels of grapefruit.

Before serving, add a healthy splash of IPA to the jug, or – if your guests are feeling brave – leave an open bottle on the table and let them top up their own drinks. Serve on the table with ice and rocks glasses.

HOW TO . . . UPCYCLE YOUR GIN BOTTLE

Sometimes the worst part of finishing a beautiful craft gin is throwing out the bottle! If you come across a bottle too gorgeous to let go, try transforming it into a work of art following these two suggestions.

A GIN LAMP

Turning your favourite gin bottle into delightful decor is as easy as buying a ready-made cork-and-fairy-light combination, so get crafty and try some DIY!

- Safety gloves and goggles
- Drill with a 10mm (½in) glass-cutting drill bit
- Bowl of cool water
- Empty, clean craft gin bottle with a cork or cap
- String of battery-operated fairy lights
- Hot-glue gun or electrical/duct tape and scissors

1. In a safe area, away from pets and children, get your safety gear on and set up your workstation. Insert your drill bit, have a bowl of water ready, and remove the cork or cap from your gin bottle and set aside.

2. Very slowly, pausing frequently to cool down your drill bit in the water, make a hole in the back of your gin bottle, towards the base. Proceed very carefully, particularly when you get close to breaking through the other side. The last thing you want is to shatter your bottle! Once your hole has been made, carefully dispose of any loose glass and wash out your bottle.

3. Slide your fairy lights through the hole and pull them up, out through the top of the bottle, until you're stopped by the battery pack. Secure the wire, at the halfway point, to the bottom of the cap or cork using hot glue or tape. If using glue, allow time to dry completely.

4. Holding the cap or cork aloft, slide the fairy lights – the strand now folded in half – back into the gin bottle. Adjust as best you can so that the lights seem to fill as much of the bottle as possible. Cork or cap your gin bottle, turn on your fairy lights, and enjoy!

A GIN VASE

Découpaging your favourite gin bottle gives it a fresh twist, and transforms it from an everyday empty to a beautiful statement piece for your home – perfect for keeping flowers in!

- Empty, clean craft gin bottle
- 10 tbsp bicarbonate of soda
- Rag
- Sponge
- Matt emulsion paint (in your choice of colour)
- Paintbrush
- Colourful paper (such as wrapping or tissue paper, photographs, vintage illustrations or pages from a book, magazine or newspaper)
- Scissors
- Glue

1. First remove the gin label. Fill a sink with warm water and submerge your bottle. Add the bicarbonate of soda and leave for 30 minutes. Rub away the label with a rag. Wait for the bottle to completely dry before moving on to the next step.

2. Using a sponge, evenly apply the matt emulsion paint. Allow the bottle to dry, and then apply a second layer.

3. While your paint is drying, cut out the coloured paper into desired patterns (e.g. flowers).

4. Use a paintbrush to apply a layer of glue to the dried, painted surface of the bottle. Apply the cut-out paper shapes, smoothing out any creases. Leave the bottle to dry for 1–2 hours.

APPLE SOUR

Another play on the classic sour, this drink is crisp and refreshing, just like autumn's best apples.

SERVES 1

MEDIUM

- 35ml (1¼fl oz) gin
- 20ml (¾fl oz) fresh lemon juice
- 1 egg white
- 2 tsp simple syrup (see page 10)
- Dry cider, to top up

Put the first four ingredients into your cocktail shaker and dry shake (without ice) to emulsify the egg. Add ice and shake again, then strain into a rocks glass. Top up with cider and enjoy!

APPLE AND CINNAMON GIMLET

Warming, spiced and sweet, this cocktail captures
the perfectly matched flavours of a comforting apple
crumble in liquid form

SERVES 1
MEDIUM

- 50ml (1¾fl oz) gin
- 1 tbsp fresh lemon juice
- Cinnamon stick and slices of apple, to garnish

FOR THE APPLE AND CINNAMON SYRUP
- 250ml (9fl oz) apple juice
- 250g (9fl oz) caster sugar
- 2 tsp ground cinnamon

Combine all the ingredients for the apple and
cinnamon syrup in a small saucepan – adding
more cinnamon if you'd like the flavour to be
stronger. Allow to bubble over a medium heat until
reduced to a syrupy consistency, then remove
from the heat and leave to cool before storing in
the fridge until needed (see also page 10).

Add the gin and lemon juice to a cocktail shaker
with 20ml (¾fl oz) of the apple and cinnamon
syrup and fill with ice. Shake and strain into a
Martini glass or cocktail coupe, then garnish with
a cinnamon stick and apple slices to serve.

PORK FILLET WITH GIN APPLE SAUCE

A classic combination that is brought to life with homemade apple sauce, using the season's best produce combined with a ginny, junipery hit.

SERVES 4–6

- 6 juniper berries
- 1 heaped tsp finely chopped fresh root ginger
- Finely grated zest of 1 lemon
- 2 tsp chopped fresh thyme leaves
- About 20 thin rashers of smoked streaky bacon
- 2 whole pork fillets (tenderloins), 500g–600g (1lb 2oz–1lb 5oz) each
- 2 tbsp vegetable oil
- 30g (1oz) unsalted butter
- 2 medium eating apples, peeled, cored and chopped
- 60ml (2fl oz) gin
- 100ml (3½fl oz) cloudy apple juice
- 3 tsp concentrated chicken stock
- Freshly ground black pepper
- Sautéed apples, to serve (see tip on page 110)

1. Use a pestle and mortar to grind the juniper berries and ginger to a smooth paste. Add the lemon zest, thyme and 2 of the bacon rashers, chopped into small pieces. Season with black pepper.

2. Take the pork fillets and trim off the ends to form a neat cylinder shape (the ends can be kept for a stir-fry). Also cut away any silvery sinew on the surface of the fillet, as this will cause the meat to curl up when cooking. Cut a slit along the length of each fillet, slicing about halfway down into the meat, then stuff the juniper, ginger, herb and bacon mixture into this slit.

3. Now take two pieces of foil, each longer and wider than the fillets. Brush the foil with some of the oil and lay strips of the remaining bacon on top, each strip slightly overlapping the other. When you have enough bacon to cover the length of the fillet, lay each fillet on top, at right angles to the bacon strips, and use the foil to wrap the bacon around the fillet. Make sure you have wrapped the pork as tightly as possible in the foil, then squeeze the ends of the foil tight like two Christmas crackers.

4. Pop the two fillets into a steamer – or in a steamer basket set over a large saucepan of simmering water and covered with a lid – and steam for 25–30 minutes, depending on size.

5. After steaming, place the foil-wrapped fillets on a plate to cool and collect any juices that run out. At this stage the parcels can be placed in the fridge overnight, or else left until cool enough to handle and unwrap. (Retain the foil for covering the pork later.)

Continues over the page

6. Preheat the oven to 200°C/180°C fan/Gas 6.
7. Melt the butter in a frying pan with the remaining oil to prevent burning and brown the pork rolls on all sides. Reserve the butter in the pan.
8. Place the pork rolls in a roasting tin and heat in the oven for around 20 minutes, or until heated through.
9. Meanwhile, add the chopped apples to the butter in the pan and cook gently for 5 minutes. Add the gin, apple juice, chicken stock and any juices collected from the pork and simmer gently until the apple is soft.
10. Place in a blender and purée until perfectly smooth. Check and adjust the seasoning. If the sauce is a little too thick for your taste, then slacken it down with a little more apple juice and/or gin.
11. Remove the pork and leave to rest, covered in foil, for 5 minutes, then carve into 2cm (¾in) slices. Serve with a little of the gin apple sauce and some sautéed apples.

TIP: To make sautéed apples, simply slice apples into rounds or wedges (no need to peel first) and brown in a pan in a little melted butter.

TWO NEGRONIS

With just three ingredients and a ratio of 1:1:1, the Negroni is a classic that's as simple as it is delicious! It pairs brilliantly with savoury snacks but if the bitterness doesn't suit your palate, the Sloe Gin version offers a mellower, sweeter alternative.

CLASSIC NEGRONI

SERVES 1

MEDIUM

· 25ml (1fl oz) gin
· 25ml (1fl oz) Campari
· 25ml (1fl oz) sweet vermouth (see tip)
· Slice or twist of orange peel, to garnish

Fill a short rocks glass with ice and pour all the ingredients on top. Stir gently for a good minute, until the drink is well chilled. Garnish with a slice or twist of orange peel.

TIP: With equal parts of all the ingredients it's important to invest in good vermouth. Cinzano Rosso, Martini Rossa or Carpano Antica are safe bets to guarantee delicious results.

SLOE GIN NEGRONI

SERVES 1

MEDIUM

· 20ml (¾fl oz) sloe gin
· 1 tsp gin
· 20ml (¾fl oz) Aperol
· 30ml (1fl oz) dry vermouth
· Twist of orange peel, to garnish

Fill a rocks glass with ice and pour over all the ingredients. Stir gently until fully chilled and serve garnished with an orange twist.

BACK TO BASICS: USING NON-LIQUIDS IN COCKTAILS

Cocktails may be for drinking, but that doesn't mean you're limited only to liquids when you're mixing one up. With a little tweak of your technique, everything from oozing honey to your favourite fruits can be incorporated.

HONEY

Honey may seem like a liquid, but it's actually too thick to be shaken into a cocktail – in fact, when combined with ice, it forms tough little clumps that won't incorporate no matter how hard you shake!

Instead, honey needs to be stirred into your cocktails before you add the ice. Add your gin to the glass or cocktail shaker, take a spoonful of honey and stir until it has completely dissolved. Only then should you add the other ingredients, finish with ice and continue making your drink.

JAM AND MARMALADE

Like honey, jams and marmalades can add an instant hit of sweetness and complexity – plus fruity flavours – into your cocktails. But they can be very stubborn.

The key to using jams and marmalades is to stir them into your spirit before you add ice to whatever drink you're making. Measure out your gin into a cocktail shaker or glass, stir

in the jam or marmalade, and then continue constructing your drink. And remember: if you're shaking, shake hard!

SWEETS

From gummy bears to hard toffees, sweets can be great to incorporate into cocktails! The best way of getting these playful flavours into your drink is to turn them into a simple syrup – head to page 10 to learn how.

FRESH FRUITS

Fresh fruits are bursting with flavour but just sticking them in a shaker won't get you far. How you approach fresh fruit will, of course, depend on what you're using. But, as a rule, you're looking at one of three techniques: juicing, muddling or making a syrup.

Juicing is pretty self-explanatory. Use a juicer or, for soft fruit, simply squeezing it through a sieve or similar, will release its juices, making it easy to use in a cocktail.

Muddling is done when you want to

release the flavours and aromas of a fruit and don't mind that the bits – the skin, seeds or larger chunks of the flesh – stay at the bottom of your drink. In fact, keeping these elements in your drink can enhance the flavour. Just drop your fruit into the bottom of a sturdy glass and, using the handle of a wooden spoon, bruise or burst the fruit until the flavours are released. Then build the rest of your cocktail on that base of flavour.

Or, for a really easy way to incorporate fruity flavours with a hint of sweetness, you can make a simple syrup! Head to page 10 to find out how.

HERBS

For a herbal twist in your drink, you can make a simple syrup (head to page 10) or rely on your handy wooden spoon to crush some flavour into your cocktail.

Soft herbs like mint and basil do well with a bit of muddling – bruise them with a wooden spoon to release their flavours and aromas, then build your cocktail on top.

For harder, woodier herbs like rosemary and thyme, a syrup infusion is usually the best way to incorporate their flavours (see page 10).

SPICED PUMPKIN PUNCH

Consider this punch your go-to Halloween concoction. It's easy, fast and serves a crowd of spooks – six, to be exact!

SERVES 6
EASY

- 300ml (10½fl oz) gin
- 600ml (1 pint) cloudy apple juice
- 90ml (3¼fl oz) pumpkin purée
- 60ml (2fl oz) honey syrup
- 30ml (1fl oz) fresh lemon juice
- ½ tsp pumpkin spice mix
- Cinnamon sticks and slices of apple, to garnish

Whack everything into a punch bowl and stir to mix. Garnish with cinnamon sticks and apple slices and serve – scarily easy!

HOW TO . . . MAKE FLAVOURED GIN AT HOME

Do you fancy a change from your usual G&T, or have a hankering for a cocktail with an extra punch of flavour? We love flavouring gin at home . . . and you'll be amazed by how easy it is to infuse gin with your favourite fruits and spices.

WHAT DO I NEED?

To make a flavoured gin at home, all you need is a sterilised jar or bottle, a good-quality gin (we like to use neutral, juniper-forward gins for making flavoured gins) and the fruits, herbs, spices and flowers you want to flavour your gin with. See below for some suggestions.

To sterilise jars and bottles, wash (along with any lids) in warm, soapy water, then rinse and dry out in the oven at a low heat (130°C/110°C fan/Gas 1) for 15 minutes. Carefully remove from the oven and use while still hot. Alternatively, place upside down in a dishwasher and run through a hot cycle.

WHAT SHOULD THE GIN-TO-FLAVOURING RATIO BE?

A good guide to follow is a third fruit to gin – in other words, about 300g (11oz) of prepared fruit for every litre (1¾ pints) of gin. Of course, the more produce or spices you use, the stronger the flavours will be, so experiment until you hit the sweet spot for your palate. Just remember that strong flavours – like citrus peel, chilli and fresh or dried herbs – can be powerful, even in small amounts.

WHAT DO I DO?

Just combine your gin and flavourings in a sterilised jar or bottle. Seal with a lid, give the mixture a good shake and place in a cool, dark place to infuse. Come back and give it a shake every couple of days, or more frequently if infusing for just a few hours (see next point). When it's done, strain into a second sterilised bottle, seal and enjoy!

HOW LONG DOES IT TAKE?

The longer you leave the gin to infuse, the stronger the flavour will be. Herbs and spices may need only a few hours, whereas veggies, fruit and berries will need up to a month to impart their flavours to the gin. But beware of leaving the infusion too long, or it will spoil! Taste the gin at regular intervals, and once you're happy with the flavour, strain out the botanicals and enjoy.

HOW SHOULD I SERVE MY FLAVOURED GIN?

Just pour over ice and add your choice of tonic and garnish! Or, if you've whipped up a batch of Toffee Gin (see page 118), why not try our Salted Toffee Martini (see page 122)?

CUCUMBER GIN

Simple and so refreshing with ice and tonic! Try adding grapefruit or lemon slices or even chilli powder to your infusion for a twist.

- 100g (3½oz) cucumber, sliced
- 500ml (18fl oz) gin

Combine the cucumber and gin in a sterilised jar or bottle, then shake and store for up to a month, shaking every couple of days.

ELDERFLOWER AND LEMON GIN

Amazing as a G&T with a little bit of ice, or as the basis of an Elderflower Tom Collins (see page 76).

- Large bunch of elderflowers, stalks trimmed off
- Several twists of lemon peel
- 500ml (18fl oz) gin

Place the elderflowers, lemon peel and gin in a sterilised jar or bottle, give the mixture a shake and leave to infuse for up to a month, shaking every couple of days.

RASPBERRY AND ROSE GIN

A romantic tipple for use in a Clover Club (see page 68) or a G&T.

- 125g (4½oz) dried rosebuds
- 125g (4½oz) fresh raspberries
- 700ml (1¼ pints) gin

Combine all the ingredients in a sterilised jar or bottle, then shake and leave to infuse for up to a month, shaking the mixture every couple of days.

RHUBARB AND GINGER GIN

Perfect with ginger beer!

- 1kg (2lb 3oz) rhubarb stalks, chopped
- 2–3 cubes peeled fresh root ginger
- 400g (14oz) caster sugar
- 800ml (1 pint 8fl oz) gin

Place all the ingredients into a sterilised jar, give the mixture a shake and store for up to a month, shaking every couple of days.

THYME GIN

A wonderfully savoury twist for a G&T or with Sicilian lemonade.

- Bunch of thyme sprigs
- 500ml (18fl oz) gin

Combine the thyme and gin in a sterilised jar or bottle, then shake and leave to infuse for a few hours, shaking the mixture every now and then.

TOFFEE GIN

Try this in our Salted Toffee Martini on page 122!

- 135g (4½oz) bag of Werther's Original toffees (the hard toffees, not the soft-centred ones)
- 200ml (7fl oz) gin

Place the toffees in a food processor and blitz into a powder. Add to a sterilised jar or bottle, pour in the gin and give the mixture a good shake. Leave to infuse for a few hours, shaking every once in a while, until the toffee is completely incorporated into the gin.

GIN-INFUSED TOMATO CHUTNEY

No Bonfire Night is complete without a warming feast of hot dogs eaten around the fire – and this delicious chutney is the perfect sausage accompaniment. It also pairs well with burgers and cold meats, and would be great dolloped on a fire-baked jacket potato filled with tangy Cheddar cheese.

MAKES 1 LARGE JAR

- 1 tbsp extra-virgin olive oil
- 1 tbsp unsalted butter
- 1 large shallot, finely chopped
- 2 juniper berries, crushed
- 250g (9oz) ripe tomatoes, chopped
- 2 tbsp soft light brown sugar
- 75ml (3fl oz) gin
- 1 tsp salt
- ¼ tsp red chilli flakes
- 1 tbsp red wine vinegar
- 1 tsp chopped fresh basil leaves

1. Heat the oil and butter over a medium-high heat until the butter has melted and is beginning to foam. Add the shallot and juniper berries and lower the heat to medium. Next, add the tomatoes and cook for 3–5 minutes. As the tomatoes start to break down, add the sugar, gin, salt and chilli flakes. Cook for another 10 minutes, pushing down the tomatoes with a wooden spoon to smash them. The sauce will begin to thicken.
2. Now add the red wine vinegar, stirring to incorporate. Continue to cook for another minute or two and then remove from the heat. Add the basil and allow to cool.
3. When cool, decant into a clean Kilner jar and store in the fridge. It will keep for a week.

SMOKY MARTINI

This recipe replaces the vermouth in a classic Martini with a dash of Scotch whisky for a tipple that is sure to warm your cockles on a cold November night! Blended Scotch is fine, but go for a peatier whisky for an even more intensely smoky bonfire hit.

SERVES 1

MEDIUM

- 40ml (1½fl oz) gin
- 25ml (1fl oz) dry vermouth
- 5ml (¼fl oz) amaretto
- 5ml (¼fl oz) Scotch whisky
- Twist of lemon peel, to garnish

Add the gin, vermouth, amaretto and whisky to a mixing glass and fill with ice. Stir gently until fully chilled and strain into a Martini or rocks glass. Garnish with a twist of lemon peel.

SALTED TOFFEE MARTINI

Forget toffee apples – it's all about toffee gin for us come Bonfire Night! Whip one of these up and it won't just be the fireworks making you 'ooh' and 'aah' with delight.

SERVES 1

EXPERT

- 40ml (1½fl oz) toffee gin (see page 118)
- 20ml (¾fl oz) chocolate liqueur
- ¾ tbsp amaretto
- Chocolate syrup and grated salted chocolate, to garnish

Add all the ingredients to a cocktail shaker, fill with ice and shake vigorously.

Pour some chocolate syrup into a saucer and dip the edge of a martini glass into the syrup. Sprinkle grated salted chocolate on to another saucer and dip the coated glass in it, so the flakes stick to the syrup, creating a chocolate rim.

Pour the contents of the shaker into your chocolatey glass and sprinkle with more grated chocolate – enjoy!

WINTER

The nights have drawn in and the world has slowed down – but that doesn't mean we have to stop having fun! Warming gin recipes, cracking Christmas cocktails and fabulously fizzy delights will have us celebrating all the way to the end of the year.

CRAFT GIN CLUB'S CHRISTMAS COCKTAIL

Let's face it: the best thing about winter is the anticipation of all that Christmas cheer. This super-easy cocktail is a staple of every Craft Gin Club Christmas party. If you have a bottle of sherry knocking about and aren't sure what to do with it, this is the Christmas cracker you've been waiting for!

SERVES 1
EASY

- 35ml (1¼fl oz) gin
- 15ml (½fl oz) sweet sherry
- 50ml (1¾fl oz) cranberry juice
- Twist of orange peel, to garnish

Fill a cocktail shaker with ice and add all your ingredients. Shake and then strain into a cocktail coupe. Garnish with a twist of orange peel and serve.

GIN AND GINGER

When the weather turns, we like to swap our tonic out for warming ginger beer. Try these interpretations of the classic Gin Mule to warm the cockles of your heart.

GIN MULE

SERVES 1

EASY

- 2 small (50p-sized) slices of fresh root ginger
- 50ml (1¾fl oz) gin
- 1 tbsp fresh lime juice
- Ginger beer, to top up
- Lime wedge, to garnish

Muddle the ginger in a cocktail shaker and combine with the gin and lime juice. Shake, then strain into a highball glass or a copper mug and fill with crushed ice. Top up with ginger beer and garnish with a lime wedge.

MARMALADE GIN MULE

SERVES 1

EASY

- 50ml (1¾fl oz) gin
- 1 heaped tsp marmalade
- 20ml (¾fl oz) fresh lime juice
- Ginger beer, to top up
- Lime wedge, to garnish

Pour the gin into a cocktail shaker and stir in your marmalade. Add the lime juice, shake and then strain into a highball glass or a copper mug packed with crushed ice. Top up with ginger beer, garnish with a lime wedge and serve.

WINTER WARMERS

Did you know that you can drink your gin warm?
These cosy cocktails are perfect for long, cold nights
– just make sure to add the gin last, so that you don't
boil away the alcohol.

GIN HOT TODDY

SERVES 1

EASY

- 75ml (3fl oz) cider
- 75ml (3fl oz) cloudy apple juice
- 2 dashes each of Angostura and orange bitters (optional)
- Pinch of ground cinnamon
- 2 tsp runny honey
- 50ml (1¾fl oz) gin
- Cinnamon stick, to garnish

Bring the cider to a simmer in a saucepan on the hob, then add the apple juice, cinnamon, bitters (if using) and honey. Stir until happily bubbling, then take off the heat. Add your gin and tip into a mug, then garnish with a cinnamon stick and serve.

GIN-GER TEA

SERVES 1

MEDIUM

- 150ml (5fl oz) freshly brewed weak black tea
- 20ml (¾fl oz) ginger syrup (see page 11)
- 20ml (¾fl oz) fresh lime juice
- 40ml (1½fl oz) gin
- Cinnamon stick, to garnish

Pour the tea into a mug and stir in the ginger syrup. Add the lime juice and tip in the gin. Stir, garnish with a cinnamon stick and serve.

WINTER G&T

Who says gin and tonic has to be a summer drink? Unsurprisingly, here at Craft Gin Club we're partial to a nice G&T all year round – and this softly spiced version is perfect for the cooler months.

SERVES 1

EASY

- 50ml (1¾fl oz) gin
- Tonic water, to taste
- Slices of orange
- Pinch of ground nutmeg
- 2 cinnamon sticks, to garnish

Add ice to a copa de balón glass or highball and pour in the gin and tonic. Add the orange slices and sprinkle the ground nutmeg on top. Garnish with the cinnamon sticks and enjoy!

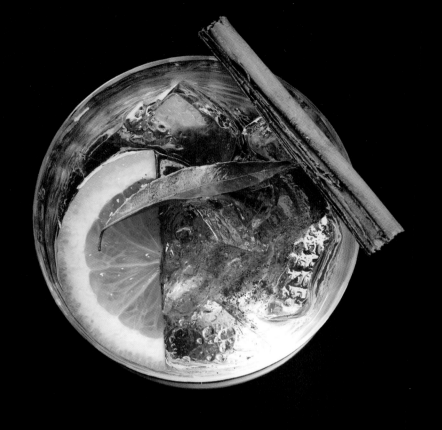
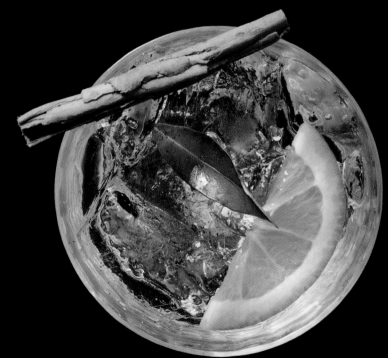

HOW TO . . . MAKE THE PERFECT GIFT FOR THE GIN LOVER IN YOUR LIFE

When the festive season rolls around, why not give your loved ones the gift of gin? Here are three cracking ideas for ginny Christmas gifts.

A BOTTLE OF CHRISTMASSY GIN LIQUEUR

Making someone a homemade clementine gin liqueur is a cheap and thoughtful alternative to a shop-bought bottle – and it couldn't be easier! Plus, if you can convince the recipient to crack it open right away, you get to enjoy it too!

MAKES 1 × 70CL–1 LITRE BOTTLE

- 1 × 70cl bottle of London Dry Gin
- 4–5 clementines
- 200g (7oz) caster sugar
- 2 star anise (optional)
- 1 small cube of fresh root ginger (optional)

YOU WILL ALSO NEED

- A decorative empty bottle, such as one used for craft gin, with a cap or cork
- Ribbon
- Gift tag

1. Leaving the skins on the clementines, slice into segments. Add to a clean, dry jar or bottle (to sterilise, see page 116), along with the other ingredients – we've suggested adding a few star anise and a little cube of unpeeled ginger for a very subtle hint of Christmas spice, but you don't have to – then seal, give the contents a little shake and leave in a cool, dark place to infuse for a week or two.

2. Give the jar a gentle shake every few days, if possible, to help the flavour infuse. You can try the contents at any point to see when you think it's ready – the longer you leave it, the deeper the flavours will be.

3. Once ready, strain your gin liqueur into a clean, pretty bottle, wrap with ribbon and tie on a gift tag. The liqueur should be kept in a cool, dark place. It will easily last for 6 months, so you can make it now and give it as a gift at Christmas without worrying about it going off.

A BOX OF GIN TRUFFLES

Shop-bought gin truffles are delicious, but making your own – using your BGF's (best gin friend's!) favourite craft gin – takes gin gifts to a whole new level.

MAKES ABOUT 12 TRUFFLES

- 400g (14oz) dark chocolate (minimum 70 per cent cocoa solids), broken into squares
- 200ml (7fl oz) double cream
- 50ml (1¾fl oz) gin
- 40ml (1½fl oz) tonic water
- Juice and grated zest of 1 lime
- Cocoa powder, to dust

YOU WILL ALSO NEED
- Gift box

1. Melt the chocolate in a bowl set over a saucepan of gently simmering water, ensuring that the bottom of the bowl doesn't touch the water. Remove from the heat, then stir in the double cream and mix until smooth.
2. Now add the gin, tonic and lime juice and zest. Mix until the combination is smooth and has thickened slightly. Transfer to an airtight container and place in the fridge to chill overnight.
3. Remove from the fridge, wash your hands and roll teaspoon-sized pieces of the mixture into balls. Place them on a tray lined with baking parchment and return to the fridge to chill for up to 2 hours.
4. Sift the cocoa powder into a bowl and roll each ball in the cocoa powder before placing in the gift box . . . or your mouth!

A MINIATURE GIN SCENT DIFFUSER

Why throw away your miniatures when you can turn them into lovely gifts for your friends? Once you've made the liquid into a nice G&T (obviously), you can add a base oil scented with essential oils and reed diffusers to the bottle for a miniature scent diffuser.

- 60ml (2fl oz) reed diffuser base oil or sweet almond oil
- 25–30 drops of a combination of essential oils of your choice (see below)
- Clean empty miniature gin bottle with a screw cap
- Funnel
- Reed diffuser sticks
- Drill
- Sticky tape

1. Pour the base oil into a glass measuring cup and add your choice of essential oils.
2. Using a funnel, pour the oil mixture into your empty gin bottle.
3. Make a hole in the cap with a drill and screw on to the bottle. Slide in the reed diffusers and secure using sticky tape.

WINTRY SCENT COMBINATIONS
- Juniper Forest: 12 drops fir needle, 8 drops pine, 6 drops cypress, 2 drop wintergreen
- Christmas Spice: 12 drops sweet orange, 6 drops cinnamon, 6 drops clove, 6 drops ginger
- Gingerbread: 10 drops ginger, 8 drops clove, 4 drops nutmeg, 4 drops cinnamon

GIN AND ORANGE GLAZED HAM JOINT

This gorgeous glazed ham makes an impressive centrepiece to any buffet table and is just the thing for feeding a crowd over the festive season. The glaze makes enough to cover a larger joint, if needed.

SERVES 10–12

- 1 × 1.5kg (3¼lb) boneless joint of unsmoked English ham
- A few strips of orange peel (minus pith)
- 6–10 juniper berries (to taste), crushed
- ½ tsp coriander seeds
- 6 cloves
- 1 cinnamon stick
- 6 allspice berries
- 4 cardamom pods

FOR THE GLAZE

- 3 tbsp shredless marmalade
- 1 tsp each ground coriander and ground nutmeg, or to taste
- 1 tbsp gin
- Thin slices of clementine (peel on)
- Cloves

1. Modern curing methods mean there is no need to soak the ham, so just pop it into a large saucepan with the orange peel and all the spices and cover with cold water. If you like your ham really spicy, or it is a large joint, you can add extra spices to taste.
2. Bring to the boil, skimming off any scum, then cover with a lid and simmer for 1½ hours (based on 30 minutes per 500g/1lb 2oz).
3. Leave to cool for 30 minutes in the cooking liquor, then remove the ham from the saucepan and drain in a colander. Cut the skin off, leaving a layer of fat, then score the fat in a diamond pattern and place in a roasting tin.
4. Now for the glaze. Place the marmalade in a small saucepan with the coriander and nutmeg, warm gently until the marmalade has melted and then add the gin. Brush the glaze over the ham and decorate with clementine slices, held in place with cloves.
5. Preheat the oven to 220°C/200°C fan/Gas 6.
6. Place the ham, uncovered, in the oven to roast for around 20 minutes, until golden and glazed on top.
7. Enjoy hot or cold with the Boozy Cranberry Relish on page 136. The cooked ham will keep, well covered, in the fridge for 10–12 days . . . if there's any left!

GIN-CURED SALMON

Gin is a fabulous curing agent, and this impressive gin-cured salmon will add a wow factor to your Christmas Day spread.

SERVES 8–10

- 400g (14oz) coarse sea salt
- 100g (3½oz) soft light brown sugar
- 150g (5oz) granulated sugar
- 1 tbsp cracked black pepper
- 1 tbsp dill seeds
- 1 tbsp coriander seeds
- 150g (5oz) fresh dill, roughly chopped
- Grated zest of 2 limes
- 100ml (3½fl oz) gin
- 1 × 1kg (2lb 3oz) piece of salmon fillet cut from the centre of the fish (or the equivalent amount of small salmon fillets), skin on and pin bones removed
- Lemon wedges, rye bread and green salad, to serve

1. Mix the salt and both sugars together in a large bowl to combine them well.
2. Use a pestle and mortar to grind the black pepper, dill and coriander seeds into a powder. Sprinkle this over the salt mixture, then stir in the chopped fresh dill, lime zest and the gin until everything is well combined.
3. Cut the salmon in half and place one piece (or half your smaller fillets, if using) skin side down on a large sheet of foil.
4. Cover the salmon with all of the salt mixture, then place the remaining salmon piece/fillets, skin side up, on top to make a sort of sandwich. Wrap the salmon tightly in the foil, then pierce a few holes in it using a cocktail stick (this allows excess liquid to drain from the fish).
5. Put the fish parcel on to a baking tray and place another baking tray on top. Weight the top baking tray down with something heavy – a couple of house bricks or weights from a set of kitchen scales, for example. Chill this package in the fridge for 2–3 days, turning the parcel of fish every 6–8 hours whenever possible.
6. When ready to serve, unwrap the salmon and scrape off any excess salt mixture. Slice the fillets very thinly on the diagonal using a sharp knife. Serve with a wedge of lemon, rye bread and a simple green salad.

BOOZY CRANBERRY RELISH

A ginny twist on a Christmas dinner table stalwart, this condiment will pair just as well with roast turkey or the Gin and Orange Glazed Ham Joint on page 134.

MAKES ABOUT 500G (1LB 2OZ)

- 400g (14oz) fresh (or frozen and defrosted) cranberries
- 75g (2½oz) golden caster sugar
- ½ tsp ground coriander
- 4 juniper berries, crushed
- Juice and grated zest of 2 clementines
- 3 tbsp gin

1. Pop all the ingredients, except the gin, into a saucepan and bring to the boil. Simmer for 5 minutes, or until the berries stop 'popping'.
2. Remove from the heat and leave to cool slightly, then stir in the gin. Serve warm or cold.

TIP: Stored in a sealable container, this relish will keep in the fridge for up to a week.

BLOOD ORANGE SOUR

Blood oranges are in season through the winter, and this cocktail makes marvellous use of them. With their beautiful, jewel-like hue, they'll look perfect served at any festive gathering.

SERVES 1

MEDIUM

- 30ml (1fl oz) gin
- 1 tbsp triple sec
- 25ml (1fl oz) fresh blood orange juice
- 1 tsp fresh lemon juice
- Slices of blood orange, to garnish

Combine all your ingredients in a cocktail shaker packed with ice. Shake and then strain into a Martini glass or cocktail coupe. Garnish with blood orange slices to serve.

THE DISTILLER'S GUIDE TO . . . JUNIPER

The cornerstone of our favourite spirit is juniper, an evergreen plant that keeps its needles even through the depths of winter. But what actually is juniper, and how does it express the terroir of where it's grown? Jason Nickels, master distiller at Salcombe Distilling Co., explains.

WHAT ACTUALLY ARE JUNIPER BERRIES?

Juniper berries are actually pine cones! They only look like berries because the scales of the cone are moulded inwards to form a round shape. They grow on low, scrubby trees or bushes and are harvested by hand in the wild. Ripe berries grow on the branch alongside unripe, green berries, so the picking technique is to rattle the branches with a stick until the ripe berries fall on to a sheet below. In a good autumn, there can be three harvests like this.

HOW CAN GIN LOVERS RECOGNISE THE TASTE OF JUNIPER IN THEIR GIN?

In gin, the dried juniper berries give the spirit a piney, resinous note – it could be considered slightly medicinal. The only official guidance on the quantity of juniper content in gin states that the predominant flavour must be juniper, but many modern gins – particularly gins with colouring added – mask this flavour with an abundance of fruit and sugar, at times even rendering it undetectable.

WHAT ARE THE MOST FAMOUS JUNIPER-PRODUCING REGIONS?

Juniper in its various forms is very widespread, and can be found from the Arctic Circle to North Africa. But nowhere is it a farmed crop – it's only ever harvested from the wild, and in most areas it just doesn't grow densely enough to make it a viable commodity. For this reason, most commercially harvested juniper comes from Macedonia and Italy, but it's also harvested in Serbia and Bosnia. In other areas, distillers may use some locally harvested crop, but rarely does this method provide enough for anything more than special editions of a gin.

HOW DOES THE FLAVOUR OF JUNIPER CHANGE DEPENDING ON WHERE IT'S GROWN?

There are a number of different species of juniper, so not only does it reflect the area in which it is grown, but the flavour can vary from one species to another. And, just like wine grapes, variations in the climate where the juniper is grown can have an effect on the moisture and oil content of the berries and, by extension, the flavour. There are a number of gins that specifically highlight these differences by using local or single varieties of juniper, while most distillers blend varieties in order to maintain a consistent flavour across generations of the same gin.

TYPES OF JUNIPER

There are hundreds of species of juniper in the world. Here are some examples of how different they can be.

Common Juniper
The most, well, common type! These berries are the blue balls of piney flavour that you'll easily recognise.

Pinchot Juniper
These berries have a Christmassy red colour rather than the bluish hue of most juniper berries, and a jammy, floral flavour.

Alligator Juniper
The bark of this variety is patterned like the animal after which it's named. It smells similar to cedar, and its berries taste strongly of vegetation.

Western Juniper
The tallest variety of juniper, its berries taste of creamy vanilla, cinnamon and even tropical fruit. One tree of this species is the Bennett Juniper, which grows in California and is 3,000 years old.

MINCE PIE MARTINI

We're mince pie obsessed here at Craft Gin Club, and this glimmering Martini always makes an appearance at our festive parties!

SERVES 1
MEDIUM

- 25ml (1fl oz) gin or sloe gin
- Sparkling white wine, to top up
- Orange peel, to garnish

FOR THE MINCEMEAT SYRUP
- 100g (3½oz) golden caster sugar
- 100ml (3½fl oz) water
- 50g (2oz) mincemeat

Combine all the ingredients for the mincemeat syrup in a saucepan and gently simmer until the mixture has reduced to a syrupy consistency. Allow to cool, then strain into an airtight container. Keep in the fridge until needed (see also page 11).

Combine the gin or sloe gin and 2 teaspoons of the mincemeat syrup in a cocktail shaker, fill with ice and shake hard. Strain into a cocktail coupe or Martini glass, top up with sparkling wine and garnish with orange peel.

UNE TRÈS BONNE ANNÉE

Named after the 75ml light field gun used by the French during the First World War, The French 75 is a classic recipe, sure to make your New Year's Eve party go with a bang! The French 77 variation adds elderflower syrup for a delicately floral sip.

FRENCH 75

SERVES 1

EASY

- 35ml (1¼fl oz) gin
- 1 tbsp fresh lemon juice
- 1 tbsp simple syrup (see page 10)
- Champagne (or sparkling white wine), to top up

Add the gin, lemon juice and syrup to a cocktail shaker and fill with ice. Shake for about 5 seconds. Strain into a champagne flute and top up with ice-cold champagne or sparkling wine for one of the best ways to kick off a dinner.

FRENCH 77

SERVES 1

EASY

- 25ml (1fl oz) gin
- 1 tbsp elderflower liqueur
- 2 tsp fresh lemon juice
- 1 tsp simple syrup (see page 10)
- 60ml (2fl oz) sparkling white wine, to top up
- Twist of lemon peel, to garnish

Combine the gin, elderflower liqueur, lemon juice and simple syrup in a cocktail shaker and fill with ice. Shake well and strain into a chilled flute or cocktail coupe. Top up with sparkling wine and garnish with a lemon twist.

SLOE GIN MINCEMEAT

Mince pies are the hallmark of Christmas! We like to mix our mincemeat with sloe gin for an extra-special treat.

**MAKES AT LEAST
1 LARGE JAR**

- 200g (7oz) raisins
- 200g (7oz) currants
- 200g (7oz) sultanas (we like to use golden sultanas)
- 100g (3½oz) dried cranberries
- 50g (2oz) chopped mixed peel
- 100g (3½oz) glacé cherries, chopped
- 100g (3½oz) blanched almonds, chopped
- 1 large Bramley apple, peeled, cored and grated
- Juice and grated zest of 1 orange
- 150ml (5fl oz) sloe gin
- 200g (7oz) soft light brown sugar
- 200g (7oz) unsalted butter, melted
- ½ tsp ground cinnamon
- ¼ tsp ground nutmeg
- ½ tsp ground mixed spice
- ¼ tsp ground allspice

1. The night before making the mincemeat, place all the dried fruit in a large bowl with the almonds, grated apple, orange juice and zest and sloe gin. Stir well and cover with cling film, then leave to soak overnight.
2. The next day add the sugar, melted butter and spices (it's a good idea to mix the spices into the sugar first, so they are evenly distributed).
3. Mix thoroughly and spoon into a large sterilised jar or jars (see page 116). As you spoon the mix in, pack it down well to avoid air pockets. Seal the jar (or jars) with a lid and store in a cool place. Best used within 6 months.

YULETIDE FLIP

Creamy, indulgent and completely delicious, this Christmassy creation is the perfect thing to sip as you deck your tree by the light of a roaring fire.

SERVES 1

MEDIUM

- 20ml (¾fl oz) gin
- 1 tbsp triple sec
- 1 tbsp Irish cream liqueur
- 1 whole egg
- Twist of orange peel, to garnish

Shake all the ingredients in a cocktail shaker, or blend in a blender with 2 ice cubes. Then serve over ice, with a twist of orange peel (Christmas baubles optional!).

HOW TO . . . THROW A GIN-TASTIC NEW YEAR'S EVE PARTY!

On New Year's Eve, the whole world is ready to party – and, after four seasons of gin-spiration, you'll be ready to host the ginniest party of the year! Here are our tips for throwing a fabulous (and gin-soaked) party to cap off the year.

STEP ONE: EMBRACE THE ROLE OF HOST

Making plans for New Year's Eve can feel like a game of chicken – everyone is waiting for their friend to make the first move. Offer to host. It's worth it. With the anticipation already running high, your guests will be ready to have a good time. That puts less pressure on you, the host, to get your friends excited for the evening. The hard work is already done!

STEP TWO: KEEP YOUR EYE ON THE CLOCK

A New Year's Eve party should be like a journey – after all, it's a long way to midnight! You don't want to lay all your cards on the table when guests arrive on the stroke of eight. From food and drink to lighting and music, start the evening with a relaxed vibe. A signature cocktail on arrival and nibbles to pick at should do the trick to start with and leaves you plenty of room to grow. As midnight approaches, turn up the music, turn down the lights and get the gin really flowing!

STEP THREE:
KEEP YOUR GUESTS IN MIND

You may be keen to crack out the gin and cut some serious shapes on the dance floor. But take a careful look at your guest list. Will everyone feel the same? Maybe some of the older guests need a cup of tea after dinner, or somewhere comfortable to sit or, if the party is in a big hall, extra heating. Depending on whether you might be going outside at any point – if you've set up an outdoor seating area, for instance, or a firepit – some guests may have thought to bring warm clothes for the evening, but others might not. Your event is tailor-made for you, but have you thought about everyone else?

STEP FOUR:
DON'T PANIC

When it comes to throwing a big party, event planners have an age-old adage that they stick to: if you fail to prepare, then prepare to fail. Give yourself lots of time to plan so you never feel flustered. Drafting in some help? Don't assume that other people know what you need them to do and be sure to communicate as clearly as possible. And, most importantly, don't stress! There will always be a problem that you didn't foresee, but equally there will always be a solution.

AND FINALLY . . .
OPEN THE GIN!

We like to start our New Year as we mean to go on: with a lot of gin! Cocktails are imperative for a good party, but why not try serving them in a new way? Whether you wheel them out on a trolley, set up a cocktail masterclass for your friends, offer a gin punch buffet or fill your bathtub with ice to keep your selection of craft gins cold, be as creative as you can. But, most importantly, have fun! As long as the music is playing, the conversation is flowing and the gin is being poured, you'll have a great party – and a great gin cocktail to toast the New Year!

POMEGRANATE GIN FIZZ

This elegant and delicious festive flute is a hit for parties. The gin and pomegranate base here can be multiplied to serve a crowd. Pre-mix it before the festivities commence and simply top up with sparkling wine in each glass as you go.

SERVES 1

EASY

- 25ml (1fl oz) gin
- 1 tsp grenadine
- Sparkling white wine, to top up
- Pomegranate seeds, to garnish

Add the gin and grenadine to a champagne flute and give a little stir. Top up with chilled sparkling wine, garnish with a few pomegranate seeds and serve immediately.

SLOE GIN FIZZ

A festive twist on an old favourite, this is perfect before (or after!) Christmas lunch. You could also substitute the champagne for soda water, for a lighter version.

SERVES 1

MEDIUM

- 30ml (1fl oz) sloe gin
- 2 tsp simple syrup (see page 10)
- 1 tbsp fresh lemon juice
- 1 egg white (optional)
- Prosecco or champagne (or sparkling wine), to top up

Add your sloe gin, syrup, lemon juice and egg white (if using; this will create a beautiful white foam on the top of your cocktail) to a cocktail shaker and shake vigorously for 30 seconds, then fill with ice and shake again. Strain into a champagne flute, then top up with Prosecco, champagne or sparkling wine.

BACK TO BASICS: THE GOLDEN RATIO

After a year of enjoying gin at home, you're more than ready to set off on your own cocktail adventures. And by applying one rule – the Golden Ratio – you can make a winner every time. It is a three-step formula that will help you create a divine cocktail without fail!

Simply combine . . .

- 50ml (1¾fl oz) spirit, such as gin
- 25ml (1fl oz) tart liquid, such as citrus juice
- 1 tbsp (15ml/½fl oz) sweet liquid, such as simple syrup (see pages 10–11) or grenadine
- A garnish

. . . and that's it!

HOW DO I COME UP WITH MY OWN COCKTAIL RECIPE?

Now that you know the Golden Ratio, all you need to do is choose flavours in each category that you think will work well together and either stir or shake them together.

Don't expect your cocktail to be a masterpiece right from the off. Depending on how inherently sweet or sour your ingredients are, you may need to adjust the proportions slightly to suit your palate. And don't be afraid to add another type of juice, a dash of sparkling mineral water or even another cordial or liqueur. Using the Golden Ratio, you won't be developing professional, perfectly calibrated drinks, but you will be able to form the basis of a new cocktail or two, made to your exact specifications!

Get the ball rolling with these super-simple combinations:

- Spirit: 50ml (1¾fl oz) gin
- Tart: 25ml (1fl oz) fresh lemon juice
- Sweet: 1 tbsp honey syrup (see page 11)
- Garnish: A lavender sprig
- Why not add . . . 1 bar spoon (teaspoon) of blackberry jam?

- Spirit: 50ml (1¾fl oz) gin
- Tart: 25ml (1fl oz) fresh orange juice
- Sweet: 1 tbsp grenadine
- Garnish: A cherry
- Why not add . . . a bit of soda water, to top up?

- Spirit: 50ml (1¾fl oz) gin
- Tart: 25ml (1fl oz) fresh lime juice
- Sweet: 1 tbsp simple syrup (see page 10)
- Garnish: A chilli lime salt rim for your cocktail glass (see page 64)
- Why not add . . . a splash of passion fruit juice?

THE LAST WORD

Sour, sweet, strong and fragrant with the herbal notes of Chartreuse, this gorgeously green classic is the perfect way to wave the old year goodbye.

SERVES 1

MEDIUM

- 20ml (¾fl oz) gin
- 20ml (¾fl oz) lime juice
- 20ml (¾fl oz) Maraschino liqueur
- 20ml (¾fl oz) green Chartreuse
- Twist of lime peel or Maraschino cherry, to garnish

Add the gin, lime juice, Maraschino liqueur and Chartreuse to a cocktail shaker and fill with ice. Shake and strain into a coupe glass. Garnish with a twist of lime peel or a Maraschino cherry.

INDEX

ACKNOWLEDGEMENTS

Here at Craft Gin Club, we've wanted to publish a book for years – and now, at long last, we've accomplished that incredible goal!

Our thanks go first and foremost to Katy Menczer, our Director of Content and Community, for spearheading this project, and Lucinda Beeman, the editor of our monthly magazine, for compiling and writing this book.

Our thanks also go out to the wonderful women who have worked so hard developing the recipes in this book: our spectacularly talented gin chef, Carol Donner; our baker extraordinaire, Eloise Benjamin; and Maria Vieira and Clementine Beach, who developed our signature serves.

Many of the ideas in this book were inspired by or adapted from the work of our in-house content team; we thank them for their creativity, hard work and spirits of limitless possibility.

And, of course, we thank the wonderful distillers who spoke with us for the Ginology sections of this book. Many thanks to Lucy and Will of Cambridge Distillery; Cameron and Stu of Four Pillars; Stuart of Shetland Distillery; and Jason of Salcombe Distilling Co.

Thank you to the wonderful team at HarperCollins: Debbie, Lydia, Harriet and Georgina for their editorial support, Sim for his spectacular design and Tom, Agathe and Becks for their work creating the wonderful pictures in this book.

But most of all, we would like to thank our amazing members, without whom none of this would be possible.

Cheers, gin lovers – have a great year of gin!